HOLT SCIENCE & TECHNOLOGY

Water on Earth

HOLT, RINEHART AND WINSTON

A Harcourt Classroom Education Company

Austin • New York • Orlando • Atlanta • San Francisco • Boston • Dallas • Toronto • London

ANNOTATED TEACHER'S EDITION

Staff Credits

Editorial

Robert W. Todd, Executive Editor

Robert V. Tucek, Leigh Ann Garcia, Senior Editors

Clay Walton, Jim Ratcliffe, Editors

Ancillaries

Jennifer Childers, Senior Editor

Chris Colby, Molly Frohlich, Shari Husain, Kristen McCardel, Sabelyn Pussman, Erin Roberson

Copyediting

Dawn Spinozza, Copyediting Supervisor

Editorial Support Staff

Jeanne Graham, Mary Helbling, Tanu'e White, Doug Rutley

Editorial Permissions

Cathy Paré, Permissions Manager

Jan Harrington, Permissions Editor

Art, Design, and Photo

Book Design

Richard Metzger, Design Director

Marc Cooper, Senior Designer

José Garza, Designer

Alicia Sullivan, Designer (ATE), **Cristina Bowerman**, Design Associate (ATE), **Eric Rupprath**, Designer (Ancillaries), **Holly Whittaker**, Traffic Coordinator

Image Acquisitions

Joe London, Director

Elaine Tate, Art Buyer Supervisor

Jeannie Taylor, Photo Research Supervisor

Andy Christiansen, Photo Researcher

Jackie Berger, Assistant Photo Researcher

Photo Studio

Sam Dudgeon, Senior Staff Photographer

Victoria Smith, Photo Specialist

Lauren Eischen, Photo Coordinator

Design New Media

Susan Michael, Design Director

Production

Mimi Stockdell, Senior Production Manager

Beth Sample, Senior Production Coordinator

Suzanne Brooks, Sara Carroll-Downs

Media Production

Kim A. Scott, Senior Production Manager

Adriana Bardin-Prestwood, Senior Production Coordinator

New Media

Armin Gutzmer, Director

Jim Bruno, Senior Project Manager

Lydia Doty, Senior Project Manager

Jessica Bega, Project Manager

Cathy Kuhles, Nina Degollado, Technical Assistants

Design Implementation and Production

The Quarasan Group, Inc.

Printed in the United States of America

ISBN 0-03-064791-6

1 2 3 4 5 6 7 048 05 04 03 02 01 00

Acknowledgments

Chapter Writers

Kathleen Meehan Berry
Science Chairman
Canon-McMillan School District
Canonsburg, Pennsylvania

Robert H. Fronk, Ph.D.
Chair of Science and Mathematics Education Department
Florida Institute of Technology
West Melbourne, Florida

Mary Kay Hemenway, Ph.D.
Research Associate and Senior Lecturer
Department of Astronomy
The University of Texas
Austin, Texas

Kathleen Kaska
Life and Earth Science Teacher
Lake Travis Middle School
Austin, Texas

Peter E. Malin, Ph.D.
Professor of Geology
Division of Earth and Ocean Sciences
Duke University
Durham, North Carolina

Karen J. Meech, Ph.D.
Associate Astronomer
Institute for Astronomy
University of Hawaii
Honolulu, Hawaii

Robert J. Sager
Chair and Professor of Earth Sciences
Pierce College
Lakewood, Washington

Lab Writers

Kenneth Creese
Science Teacher
White Mountain Junior High School
Rock Springs, Wyoming

Linda A. Culp
Science Teacher and Dept. Chair
Thorndale High School
Thorndale, Texas

Bruce M. Jones
Science Teacher and Dept. Chair
The Blake School
Minneapolis, Minnesota

Shannon Miller
Science and Math Teacher
Llano Junior High School
Llano, Texas

Robert Stephen Ricks
Special Services Teacher
Department of Classroom Improvement
Alabama State Department of Education
Montgomery, Alabama

James J. Secosky
Science Teacher
Bloomfield Central School
Bloomfield, New York

Academic Reviewers

Mead Allison, Ph.D.
Assistant Professor of Oceanography
Texas A&M University
Galveston, Texas

Alissa Arp, Ph.D.
Director and Professor of Environmental Studies
Romberg Tiburon Center
San Francisco State University
Tiburon, California

Paul D. Asimow, Ph.D.
Assistant Professor of Geology and Geochemistry
Department of Physics and Planetary Sciences
California Institute of Technology
Pasadena, California

G. Fritz Benedict, Ph.D.
Senior Research Scientist and Astronomer
McDonald Observatory
The University of Texas
Austin, Texas

Russell M. Brengelman, Ph.D.
Professor of Physics
Morehead State University
Morehead, Kentucky

John A. Brockhaus, Ph.D.
Director—Mapping, Charting, and Geodesy Program
Department of Geography and Environmental Engineering
United States Military Academy
West Point, New York

Michael Brown, Ph.D.
Assistant Professor of Planetary Astronomy
Department of Physics and Astronomy
California Institute of Technology
Pasadena, California

Wesley N. Colley, Ph.D.
Postdoctoral Fellow
Harvard-Smithsonian Center for Astrophysics
Cambridge, Massachusetts

Andrew J. Davis, Ph.D.
Manager—ACE Science Data Center
Physics Department
California Institute of Technology
Pasadena, California

Peter E. Demmin, Ed.D.
Former Science Teacher and Department Chair
Amherst Central High School
Amherst, New York

James Denbow, Ph.D.
Associate Professor
Department of Anthropology
The University of Texas
Austin, Texas

Roy W. Hann, Jr., Ph.D.
Professor of Civil Engineering
Texas A&M University
College Station, Texas

Frederick R. Heck, Ph.D.
Professor of Geology
Ferris State University
Big Rapids, Michigan

Richard Hey, Ph.D.
Professor of Geophysics
Hawaii Institute of Geophysics and Planetology
University of Hawaii
Honolulu, Hawaii

John E. Hoover, Ph.D.
Associate Professor of Biology
Millersville University
Millersville, Pennsylvania

Robert W. Houghton, Ph.D.
Senior Staff Associate
Lamont-Doherty Earth Observatory
Columbia University
Palisades, New York

Steven A. Jennings, Ph.D.
Assistant Professor
Department of Geography & Environmental Studies
University of Colorado
Colorado Springs, Colorado

Eric L. Johnson, Ph.D.
Assistant Professor of Geology
Central Michigan University
Mount Pleasant, Michigan

John Kermond, Ph.D.
Visiting Scientist
NOAA–Office of Global Programs
Silver Spring, Maryland

Zavareh Kothavala, Ph.D.
Postdoctoral Associate Scientist
Department of Geology and Geophysics
Yale University
New Haven, Connecticut

Karen Kwitter, Ph.D.
Ebenezer Fitch Professor of Astronomy
Williams College
Williamstown, Massachusetts

Valerie Lang, Ph.D.
Project Leader of Environmental Programs
The Aerospace Corporation
Los Angeles, California

Philip LaRoe
Professor
Helena College of Technology
Helena, Montana

Julie Lutz, Ph.D.
Astronomy Program
Washington State University
Pullman, Washington

Duane F. Marble, Ph.D.
Professor Emeritus
Department of Geography and Natural Resources
Ohio State University
Columbus, Ohio

Joseph A. McClure, Ph.D.
Associate Professor
Department of Physics
Georgetown University
Washington, D.C.

Frank K. McKinney, Ph.D.
Professor of Geology
Appalachian State University
Boone, North Carolina

Joann Mossa, Ph.D.
Associate Professor of Geography
University of Florida
Gainesville, Florida

LaMoine L. Motz, Ph.D.
Coordinator of Science Education
Department of Learning Services
Oakland County Schools
Waterford, Michigan

Barbara Murck, Ph.D.
Assistant Professor of Earth Science
Erindale College
University of Toronto
Mississauga, Ontario
CANADA

Hilary Clement Olson, Ph.D.
Research Associate
Institute for Geophysics
The University of Texas
Austin, Texas

Andre Potochnik
Geologist
Grand Canyon Field Institute
Flagstaff, Arizona

John R. Reid, Ph.D.
Professor Emeritus
Department of Geology and Geological Engineering
University of North Dakota
Grand Forks, North Dakota

Gary Rottman, Ph.D.
Associate Director
Laboratory for Atmosphere and Space Physics
University of Colorado
Boulder, Colorado

Dork L. Sahagian, Ph.D.
Professor
Institute for the Study of Earth, Oceans, and Space
University of New Hampshire
Durham, New Hampshire

Peter Sheridan, Ph.D.
Professor of Chemistry
Colgate University
Hamilton, New York

David Sprayberry, Ph.D.
Assistant Director for Observing Support
W.M. Keck Observatory
California Association for Research in Astronomy
Kamuela, Hawaii

Lynne Talley, Ph.D.
Professor
Scripps Institution of Oceanography
University of California
La Jolla, California

Acknowledgments (cont.)

Glenn Thompson, Ph.D.
Scientist
Geophysical Institute
University of Alaska
Fairbanks, Alaska

Martin VanDyke, Ph.D.
Professor of Chemistry Emeritus
Front Range Community College
Westminister, Colorado

Thad A. Wasklewicz, Ph.D.
Assistant Professor of Geography
University of Memphis
Memphis, Tennessee

Hans Rudolf Wenk, Ph.D.
Professor of Geology and Geophysical Sciences
University of California
Berkeley, California

Lisa D. White, Ph.D.
Associate Professor of Geosciences
San Francisco State University
San Francisco, California

Lorraine W. Wolf, Ph.D.
Associate Professor of Geology
Auburn University
Auburn, Alabama

Charles A. Wood, Ph.D.
Chairman and Professor of Space Studies
University of North Dakota
Grand Forks, North Dakota

Safety Reviewer

Jack Gerlovich, Ph.D.
Associate Professor
School of Education
Drake University
Des Moines, Iowa

Teacher Reviewers

Barry L. Bishop
Science Teacher and Dept. Chair
San Rafael Junior High School
Ferron, Utah

Yvonne Brannum
Science Teacher and Dept. Chair
Hine Junior High School
Washington, D.C.

Daniel L. Bugenhagen
Science Teacher and Dept. Chair
Yutan Junior & Senior High School
Yutan, Nebraska

Kenneth Creese
Science Teacher
White Mountain Junior High School
Rock Springs, Wyoming

Linda A. Culp
Science Teacher and Dept. Chair
Thorndale High School
Thorndale, Texas

Alonda Droege
Science Teacher
Pioneer Middle School
Steilacom, Washington

Laura Fleet
Science Teacher
Alice B. Landrum Middle School
Ponte Vedra Beach, Florida

Susan Gorman
Science Teacher
Northridge Middle School
North Richland Hills, Texas

C. John Graves
Science Teacher
Monforton Middle School
Bozeman, Montana

Janel Guse
Science Teacher and Dept. Chair
West Central Middle School
Hartford, South Dakota

Gary Habeeb
Science Mentor
Sierra–Plumas Joint Unified School District
Downieville, California

Dennis Hanson
Science Teacher and Dept. Chair
Big Bear Middle School
Big Bear Lake, California

Norman E. Holcomb
Science Teacher
Marion Local Schools
Maria Stein, Ohio

Tracy Jahn
Science Teacher
Berkshire Junior-Senior High School
Canaan, New York

David D. Jones
Science Teacher
Andrew Jackson Middle School
Cross Lanes, West Virginia

Howard A. Knodle
Science Teacher
Belvidere High School
Belvidere, Illinois

Michael E. Kral
Science Teacher
West Hardin Middle School
Cecilia, Kentucky

Kathy LaRoe
Science Teacher
East Valley Middle School
East Helena, Montana

Scott Mandel, Ph.D.
Director and Educational Consultant
Teachers Helping Teachers
Los Angeles, California

Kathy McKee
Science Teacher
Hoyt Middle School
Des Moines, Iowa

Michael Minium
Vice President of Program Development
United States Orienteering Federation
Forest Park, Georgia

Jan Nelson
Science Teacher
East Valley Middle School
East Helena, Montana

Dwight C. Patton
Science Teacher
Carroll T. Welch Middle School
Horizon City, Texas

Joseph Price
Chairman—Science Department
H. M. Brown Junior High School
Washington, D.C.

Terry J. Rakes
Science Teacher
Elmwood Junior High School
Rogers, Arkansas

Steven Ramig
Science Teacher
West Point High School
West Point, Nebraska

Helen P. Schiller
Science Teacher
Northwood Middle School
Taylors, South Carolina

Bert J. Sherwood
Science Teacher
Socorro Middle School
El Paso, Texas

Larry Tackett
Science Teacher and Dept. Chair
Andrew Jackson Middle School
Cross Lanes, West Virginia

Walter Woolbaugh
Science Teacher
Manhattan Junior High School
Manhattan, Montana

Alexis S. Wright
Middle School Science Coordinator
Rye Country Day School
Rye, New York

Gordon Zibelman
Science Teacher
Drexel Hill Middle School
Drexel Hill, Pennsylvania

Water on Earth

H

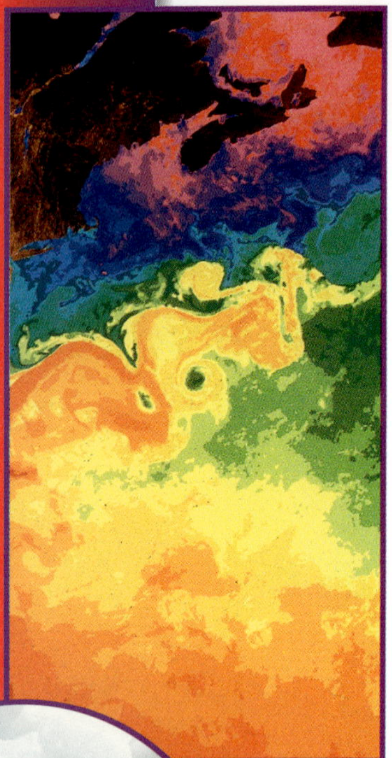

Skills Development

Process Skills

QuickLabs

Chapter Labs

Research and Critical Thinking Skills

Apply

Feature Articles

Connections

Mathematics

Program Scope and Sequence

Selecting the right books for your course is easy. Just review the topics presented in each book to determine the best match to your district curriculum.

	A MICROORGANISMS, FUNGI, AND PLANTS	B ANIMALS
CHAPTER 1	**It's Alive!! Or, Is It?** ❑ Characteristics of living things ❑ Homeostasis ❑ Heredity and DNA ❑ Producers, consumers, and decomposers ❑ Biomolecules	**Animals and Behavior** ❑ Characteristics of animals ❑ Classification of animals ❑ Animal behavior ❑ Hibernation and estivation ❑ The biological clock ❑ Animal communication ❑ Living in groups
CHAPTER 2	**Bacteria and Viruses** ❑ Binary fission ❑ Characteristics of bacteria ❑ Nitrogen-fixing bacteria ❑ Antibiotics ❑ Pathogenic bacteria ❑ Characteristics of viruses ❑ Lytic cycle	**Invertebrates** ❑ General characteristics of invertebrates ❑ Types of symmetry ❑ Characteristics of sponges, cnidarians, arthropods, and echinoderms ❑ Flatworms versus roundworms ❑ Types of circulatory systems
CHAPTER 3	**Protists and Fungi** ❑ Characteristics of protists ❑ Types of algae ❑ Types of protozoa ❑ Protist reproduction ❑ Characteristics of fungi and lichens	**Fishes, Amphibians, and Reptiles** ❑ Characteristics of vertebrates ❑ Structure and kinds of fishes ❑ Development of lungs ❑ Structure and kinds of amphibians and reptiles ❑ Function of the amniotic egg
CHAPTER 4	**Introduction to Plants** ❑ Characteristics of plants and seeds ❑ Reproduction and classification ❑ Angiosperms versus gymnosperms ❑ Monocots versus dicots ❑ Structure and functions of roots, stems, leaves, and flowers	**Birds and Mammals** ❑ Structure and kinds of birds ❑ Types of feathers ❑ Adaptations for flight ❑ Structure and kinds of mammals ❑ Function of the placenta
CHAPTER 5	**Plant Processes** ❑ Pollination and fertilization ❑ Dormancy ❑ Photosynthesis ❑ Plant tropisms ❑ Seasonal responses of plants	
CHAPTER 6		
CHAPTER 7		

Life Science

C CELLS, HEREDITY, & CLASSIFICATION	D HUMAN BODY SYSTEMS & HEALTH	E ENVIRONMENTAL SCIENCE
Cells: The Basic Units of Life ❑ Cells, tissues, and organs ❑ Populations, communities, and ecosystems ❑ Cell theory ❑ Surface-to-volume ratio ❑ Prokaryotic versus eukaryotic cells ❑ Cell organelles	**Body Organization and Structure** ❑ Homeostasis ❑ Types of tissue ❑ Organ systems ❑ Structure and function of the skeletal system, muscular system, and integumentary system	**Interactions of Living Things** ❑ Biotic versus abiotic parts of the environment ❑ Producers, consumers, and decomposers ❑ Food chains and food webs ❑ Factors limiting population growth ❑ Predator-prey relationships ❑ Symbiosis and coevolution
The Cell in Action ❑ Diffusion and osmosis ❑ Passive versus active transport ❑ Endocytosis versus exocytosis ❑ Photosynthesis ❑ Cellular respiration and fermentation ❑ Cell cycle	**Circulation and Respiration** ❑ Structure and function of the cardiovascular system, lymphatic system, and respiratory system ❑ Respiratory disorders	**Cycles in Nature** ❑ Water cycle ❑ Carbon cycle ❑ Nitrogen cycle ❑ Ecological succession
Heredity ❑ Dominant versus recessive traits ❑ Genes and alleles ❑ Genotype, phenotype, the Punnett square and probability ❑ Meiosis ❑ Determination of sex	**The Digestive and Urinary Systems** ❑ Structure and function of the digestive system ❑ Structure and function of the urinary system	**The Earth's Ecosystems** ❑ Kinds of land and water biomes ❑ Marine ecosystems ❑ Freshwater ecosystems
Genes and Gene Technology ❑ Structure of DNA ❑ Protein synthesis ❑ Mutations ❑ Heredity disorders and genetic counseling	**Communication and Control** ❑ Structure and function of the nervous system and endocrine system ❑ The senses ❑ Structure and function of the eye and ear	**Environmental Problems and Solutions** ❑ Types of pollutants ❑ Types of resources ❑ Conservation practices ❑ Species protection
The Evolution of Living Things ❑ Adaptations and species ❑ Evidence for evolution ❑ Darwin's work and natural selection ❑ Formation of new species	**Reproduction and Development** ❑ Asexual versus sexual reproduction ❑ Internal versus external fertilization ❑ Structure and function of the human male and female reproductive systems ❑ Fertilization, placental development, and embryo growth ❑ Stages of human life	**Energy Resources** ❑ Types of resources ❑ Energy resources and pollution ❑ Alternative energy resources
The History of Life on Earth ❑ Geologic time scale and extinctions ❑ Plate tectonics ❑ Human evolution	**Body Defenses and Disease** ❑ Types of diseases ❑ Vaccines and immunity ❑ Structure and function of the immune system ❑ Autoimmune diseases, cancer, and AIDS	
Classification ❑ Levels of classification ❑ Cladistic diagrams ❑ Dichotomous keys ❑ Characteristics of the six kingdoms	**Staying Healthy** ❑ Nutrition and reading food labels ❑ Alcohol and drug effects on the body ❑ Hygiene, exercise, and first aid	

Scope and Sequence *(continued)*

	F INSIDE THE RESTLESS EARTH	**G EARTH'S CHANGING SURFACE**
CHAPTER 1	**Minerals of the Earth's Crust** ❏ Mineral composition and structure ❏ Types of minerals ❏ Mineral identification ❏ Mineral formation and mining	**Maps as Models of the Earth** ❏ Structure of a map ❏ Cardinal directions ❏ Latitude, longitude, and the equator ❏ Magnetic declination and true north ❏ Types of projections ❏ Aerial photographs ❏ Remote sensing ❏ Topographic maps
CHAPTER 2	**Rocks: Mineral Mixtures** ❏ Rock cycle and types of rocks ❏ Rock classification ❏ Characteristics of igneous, sedimentary, and metamorphic rocks	**Weathering and Soil Formation** ❏ Types of weathering ❏ Factors affecting the rate of weathering ❏ Composition of soil ❏ Soil conservation and erosion prevention
CHAPTER 3	**The Rock and Fossil Record** ❏ Uniformitarianism versus catastrophism ❏ Superposition ❏ The geologic column and unconformities ❏ Absolute dating and radiometric dating ❏ Characteristics and types of fossils ❏ Geologic time scale	**Agents of Erosion and Deposition** ❏ Shoreline erosion and deposition ❏ Wind erosion and deposition ❏ Erosion and deposition by ice ❏ Gravity's effect on erosion and deposition
CHAPTER 4	**Plate Tectonics** ❏ Structure of the Earth ❏ Continental drifts and sea floor spreading ❏ Plate tectonics theory ❏ Types of boundaries ❏ Types of crust deformities	
CHAPTER 5	**Earthquakes** ❏ Seismology ❏ Features of earthquakes ❏ P and S waves ❏ Gap hypothesis ❏ Earthquake safety	
CHAPTER 6	**Volcanoes** ❏ Types of volcanoes and eruptions ❏ Types of lava and pyroclastic material ❏ Craters versus calderas ❏ Sites and conditions for volcano formation ❏ Predicting eruptions	

Earth Science

H WATER ON EARTH

The Flow of Fresh Water
- ❏ Water cycle
- ❏ River systems
- ❏ Stream erosion
- ❏ Life cycle of rivers
- ❏ Deposition
- ❏ Aquifers, springs, and wells
- ❏ Ground water
- ❏ Water treatment and pollution

Exploring the Oceans
- ❏ Properties and characteristics of the oceans
- ❏ Features of the ocean floor
- ❏ Ocean ecology
- ❏ Ocean resources and pollution

The Movement of Ocean Water
- ❏ Types of currents
- ❏ Characteristics of waves
- ❏ Types of ocean waves
- ❏ Tides

I WEATHER AND CLIMATE

The Atmosphere
- ❏ Structure of the atmosphere
- ❏ Air pressure
- ❏ Radiation, convection, and conduction
- ❏ Greenhouse effect and global warming
- ❏ Characteristics of winds
- ❏ Types of winds
- ❏ Air pollution

Understanding Weather
- ❏ Water cycle
- ❏ Humidity
- ❏ Types of clouds
- ❏ Types of precipitation
- ❏ Air masses and fronts
- ❏ Storms, tornadoes, and hurricanes
- ❏ Weather forecasting
- ❏ Weather maps

Climate
- ❏ Weather versus climate
- ❏ Seasons and latitude
- ❏ Prevailing winds
- ❏ Earth's biomes
- ❏ Earth's climate zones
- ❏ Ice ages
- ❏ Global warming
- ❏ Greenhouse effect

J ASTRONOMY

Observing the Sky
- ❏ Astronomy
- ❏ Keeping time
- ❏ Mapping the stars
- ❏ Scales of the universe
- ❏ Types of telescope
- ❏ Radioastronomy

Formation of the Solar System
- ❏ Birth of the solar system
- ❏ Planetary motion
- ❏ Newton's Law of Universal Gravitation
- ❏ Structure of the sun
- ❏ Fusion
- ❏ Earth's structure and atmosphere

A Family of Planets
- ❏ Properties and characteristics of the planets
- ❏ Properties and characteristics of moons
- ❏ Comets, asteroids, and meteoroids

The Universe Beyond
- ❏ Composition of stars
- ❏ Classification of stars
- ❏ Star brightness, distance, and motions
- ❏ H-R diagram
- ❏ Life cycle of stars
- ❏ Types of galaxies
- ❏ Theories on the formation of the universe

Exploring Space
- ❏ Rocketry and artificial satellites
- ❏ Types of Earth orbit
- ❏ Space probes and space exploration

Scope and Sequence *(continued)*

	K INTRODUCTION TO MATTER	L INTERACTIONS OF MATTER
CHAPTER 1	**The Properties of Matter** ❏ Definition of matter ❏ Mass and weight ❏ Physical and chemical properties ❏ Physical and chemical change ❏ Density	**Chemical Bonding** ❏ Types of chemical bonds ❏ Valence electrons ❏ Ions versus molecules ❏ Crystal lattice
CHAPTER 2	**States of Matter** ❏ States of matter and their properties ❏ Boyle's and Charles's laws ❏ Changes of state	**Chemical Reactions** ❏ Writing chemical formulas and equations ❏ Law of conservation of mass ❏ Types of reactions ❏ Endothermic versus exothermic reactions ❏ Law of conservation of energy ❏ Activation energy ❏ Catalysts and inhibitors
CHAPTER 3	**Elements, Compounds, and Mixtures** ❏ Elements and compounds ❏ Metals, nonmetals, and metalloids (semiconductors) ❏ Properties of mixtures ❏ Properties of solutions, suspensions, and colloids	**Chemical Compounds** ❏ Ionic versus covalent compounds ❏ Acids, bases, and salts ❏ pH ❏ Organic compounds ❏ Biomolecules
CHAPTER 4	**Introduction to Atoms** ❏ Atomic theory ❏ Atomic model and structure ❏ Isotopes ❏ Atomic mass and mass number	**Atomic Energy** ❏ Properties of radioactive substances ❏ Types of decay ❏ Half-life ❏ Fission, fusion, and chain reactions
CHAPTER 5	**The Periodic Table** ❏ Structure of the periodic table ❏ Periodic law ❏ Properties of alkali metals, alkaline-earth metals, halogens, and noble gases	
CHAPTER 6		

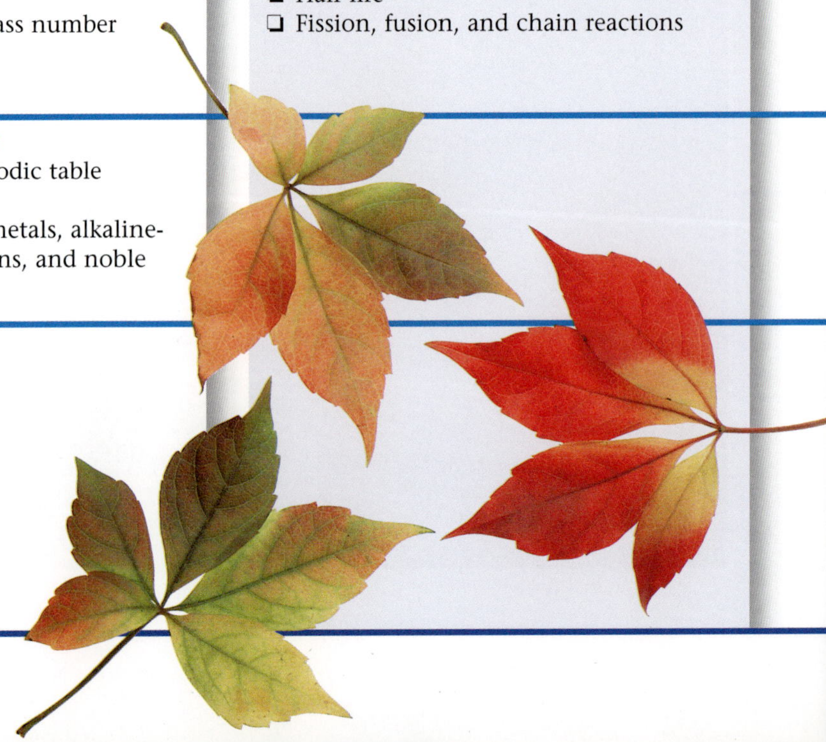

Physical Science

M FORCES, MOTION, AND ENERGY

Matter in Motion
- ❏ Speed, velocity, and acceleration
- ❏ Measuring force
- ❏ Friction
- ❏ Mass versus weight

Forces in Motion
- ❏ Terminal velocity and free fall
- ❏ Projectile motion
- ❏ Inertia
- ❏ Momentum

Forces in Fluids
- ❏ Properties in fluids
- ❏ Atmospheric pressure
- ❏ Density
- ❏ Pascal's principle
- ❏ Buoyant force
- ❏ Archimedes' principle
- ❏ Bernoulli's principle

Work and Machines
- ❏ Measuring work
- ❏ Measuring power
- ❏ Types of machines
- ❏ Mechanical advantage
- ❏ Mechanical efficiency

Energy and Energy Resources
- ❏ Forms of energy
- ❏ Energy conversions
- ❏ Law of conservation of energy
- ❏ Energy resources

Heat and Heat Technology
- ❏ Heat versus temperature
- ❏ Thermal expansion
- ❏ Absolute zero
- ❏ Conduction, convection, radiation
- ❏ Conductors versus insulators
- ❏ Specific heat capacity
- ❏ Changes of state
- ❏ Heat engines
- ❏ Thermal pollution

N ELECTRICITY AND MAGNETISM

Introduction to Electricity
- ❏ Law of electric charges
- ❏ Conduction versus induction
- ❏ Static electricity
- ❏ Potential difference
- ❏ Cells, batteries, and photocells
- ❏ Thermocouples
- ❏ Voltage, current, and resistance
- ❏ Electric power
- ❏ Types of circuits

Electromagnetism
- ❏ Properties of magnets
- ❏ Magnetic force
- ❏ Electromagnetism
- ❏ Solenoids and electric motors
- ❏ Electromagnetic induction
- ❏ Generators and transformers

Electronic Technology
- ❏ Properties of semiconductors
- ❏ Integrated circuits
- ❏ Diodes and transistors
- ❏ Analog versus digital signals
- ❏ Microprocessors
- ❏ Features of computers

O SOUND AND LIGHT

The Energy of Waves
- ❏ Properties of waves
- ❏ Types of waves
- ❏ Reflection and refraction
- ❏ Diffraction and interference
- ❏ Standing waves and resonance

The Nature of Sound
- ❏ Properties of sound waves
- ❏ Structure of the human ear
- ❏ Pitch and the Doppler effect
- ❏ Infrasonic versus ultrasonic sound
- ❏ Sound reflection and echolocation
- ❏ Sound barrier
- ❏ Interference, resonance, diffraction, and standing waves
- ❏ Sound quality of instruments

The Nature of Light
- ❏ Electromagnetic waves
- ❏ Electromagnetic spectrum
- ❏ Law of reflection
- ❏ Absorption and scattering
- ❏ Reflection and refraction
- ❏ Diffraction and interference

Light and Our World
- ❏ Luminosity
- ❏ Types of lighting
- ❏ Types of mirrors and lenses
- ❏ Focal point
- ❏ Structure of the human eye
- ❏ Lasers and holograms

Holt Science & Technology

Components Listing

Effective planning starts with all the resources you need in an easy-to-use package for each short course.

Directed Reading Worksheets Help students develop and practice fundamental reading comprehension skills and provide a comprehensive review tool for students to use when studying for an exam.

Study Guide Vocabulary & Notes Worksheets and Chapter Review Worksheets are reproductions of the Chapter Highlights and Chapter Review sections that follow each chapter in the textbook.

Science Puzzlers, Twisters & Teasers Use vocabulary and concepts from each chapter of the Pupil's Editions as elements of rebuses, anagrams, logic puzzles, daffy definitions, riddle poems, word jumbles, and other types of puzzles.

Reinforcement and Vocabulary Review Worksheets Approach a chapter topic from a different angle with an emphasis on different learning modalities to help students that are frustrated by traditional methods.

Critical Thinking & Problem Solving Worksheets Develop the following skills: distinguishing fact from opinion, predicting consequences, analyzing information, and drawing conclusions. Problem Solving Worksheets develop a step-by-step process of problem analysis including gathering information, asking critical questions, identifying alternatives, and making comparisons.

Math Skills for Science Worksheets Each activity gives a brief introduction to a relevant math skill, a step-by-step explanation of the math process, one or more example problems, and a variety of practice problems.

Science Skills Worksheets Help your students focus specifically on skills such as measuring, graphing, using logic, understanding statistics, organizing research papers, and critical thinking options.

LAB ACTIVITIES

ALL LABS ARE CLASSROOM TESTED & APPROVED

Datasheets for Labs These worksheets are the labs found in the *Holt Science & Technology* textbook. Charts, tables, and graphs are included to make data collection and analysis easier, and space is provided to write observations and conclusions.

Whiz-Bang Demonstrations Discovery or Making Models experiences label each demo as one in which students discover an answer or use a scientific model.

Calculator-Based Labs Give students the opportunity to use graphing-calculator probes and sensors to collect data using a TI graphing calculator, Vernier sensors, and a TI CBL 2™ or Vernier Lab Pro interface.

EcoLabs and Field Activities Focus on educational outdoor projects, such as wildlife observation, nature surveys, or natural history.

Inquiry Labs Use the scientific method to help students find their own path in solving a real-world problem.

Long-Term Projects and Research Ideas Provide students with the opportunity to go beyond library and Internet resources to explore science topics.

ASSESSMENT

Chapter Tests Each four-page chapter test consists of a variety of item types including Multiple Choice, Using Vocabulary, Short Answer, Critical Thinking, Math in Science, Interpreting Graphics, and Concept Mapping.

Performance-Based Assessments Evaluate students' abilities to solve problems using the tools, equipment, and techniques of science. Rubrics included for each assessment make it easy to evaluate student performance.

TEACHER RESOURCES

Lesson Plans Integrate all of the great resources in the *Holt Science & Technology* program into your daily teaching. Each lesson plan includes a correlation of the lesson activities to the National Science Education Standards.

Teaching Transparencies Each transparency is correlated to a particular lesson in the Chapter Organizer.

Concept Mapping Transparencies, Worksheets, and Answer Key
Give students an opportunity to complete their own concept maps to study the concepts within each chapter and form logical connections. Student worksheets contain a blank concept map with linking phrases and a list of terms to be used by the student to complete the map.

TECHNOLOGY RESOURCES

One-Stop Planner CD-ROM
Finding the right resources is easy with the One-Stop Planner CD-ROM. You can view and print any resource with just the click of a mouse. Customize the suggested lesson plans to match your daily or weekly calendar and your district's requirements. Powerful test generator software allows you to create customized assessments using a databank of items.

The One-Stop Planner for each level includes the following:

- All materials from the Teaching Resources
- Bellringer Transparency Masters
- Block Scheduling Tools
- Standards Correlations
- Lab Inventory Checklist
- Safety Information
- Science Fair Guide
- Parent Involvement Tools
- Spanish Audio Scripts
- Spanish Glossary
- Assessment Item Listing
- Assessment Checklists and Rubrics
- Test Generator

*sci*LINKS
*sci*LINKS numbers throughout the text take you and your students to some of the best on-line resources available. Sites are constantly reviewed and updated by the National Science Teachers Association. Special "teacher only" sites are available to you once you register with the service.

go.hrw.com
To access Holt, Rinehart and Winston Web resources, use the home page codes for each level found on page 1 of the Pupil's Editions. The codes shown on the Chapter Organizers for each chapter in the Annotated Teacher's Edition take you to chapter-specific resources.

Smithsonian Institution
Find lesson plans, activities, interviews, virtual exhibits, and just general information on a wide variety of topics relevant to middle school science.

CNNfyi.com
Find the latest in late-breaking science news for students. Featured news stories are supported with lesson plans and activities.

CNN Presents Science in the News Video Library
Bring relevant science news stories into the classroom. Each video comes with a Teacher's Guide and set of Critical Thinking Worksheets that develop listening and media analysis skills. Tapes in the series include:

- Eye on the Environment
- Multicultural Connections
- Scientists in Action
- Science, Technology & Society

Guided Reading Audio CD Program
Students can listen to a direct read of each chapter and follow along in the text. Use the program as a content bridge for struggling readers and students for whom English is not their native language.

Interactive Explorations CD-ROM
Turn a computer into a virtual laboratory. Students act as lab assistants helping Dr. Crystal Labcoat solve real-world problems. Activities develop students' inquiry, analysis, and decision-making skills.

Interactive Science Encyclopedia CD-ROM
Give your students access to more than 3,000 cross-referenced scientific definitions, in-depth articles, science fair project ideas, activities, and more.

ADDITIONAL COMPONENTS

Holt Anthology of Science Fiction
Science Fiction features in the Pupil's Edition preview the stories found in the anthology. Each story begins with a Reading Prep guide and closes with Think About It questions.

Professional Reference for Teachers
Articles written by leading educators help you learn more about the National Science Education Standards, block scheduling, classroom management techniques, and more. A bibliography of professional references is included.

Holt Science Posters
Seven wall posters highlight interesting topics, such as the Physics of Sports, or useful reference material, such as the Scientific Method.

Holt Science Skills Workshop: Reading in the Content Area
Use a variety of in-depth skills exercises to help students learn to read science materials strategically.

Key

These materials are blackline masters.

All titles shown in green are found in the *Teaching Resources* booklets for each course.

Science & Math Skills Worksheets

The *Holt Science and Technology* program helps you meet the needs of a wide variety of students, regardless of their skill level. The following pages provide examples of the worksheets available to improve your students' science and math skills, whether they already have a strong science and math background or are weak in these areas. Samples of assessment checklists and rubrics are also provided.

In addition to the skills worksheets represented here, *Holt Science and Technology* provides a variety of worksheets that are correlated directly with each chapter of the program. Representations of these worksheets are found at the beginning of each chapter in this Annotated Teacher's Edition. Specific worksheets related to each chapter are listed in the Chapter Organizer. Worksheets and transparencies are found in the softcover *Teaching Resources* for each course.

Many worksheets are also available on the HRW Web site. The address is **go.hrw.com.**

Science Skills Worksheets: Thinking Skills

BEING FLEXIBLE

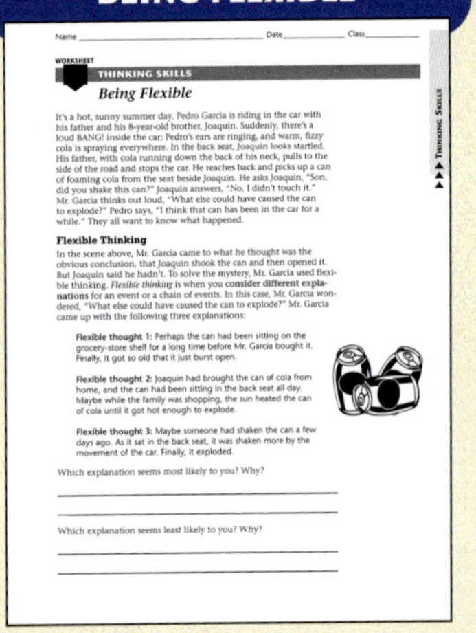
THINKING SKILLS

Being Flexible

USING YOUR SENSES

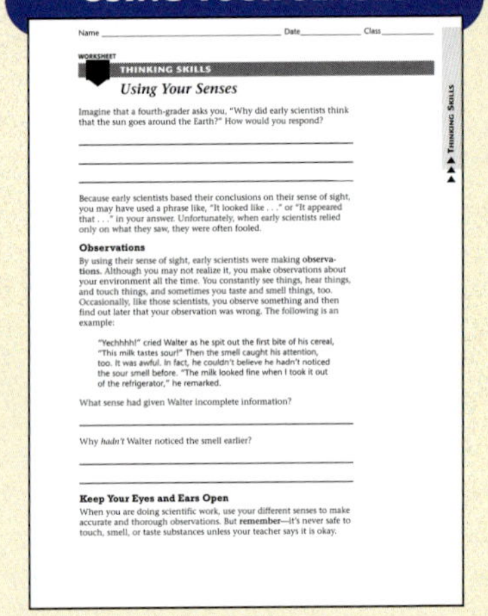
THINKING SKILLS

Using Your Senses

THINKING OBJECTIVELY

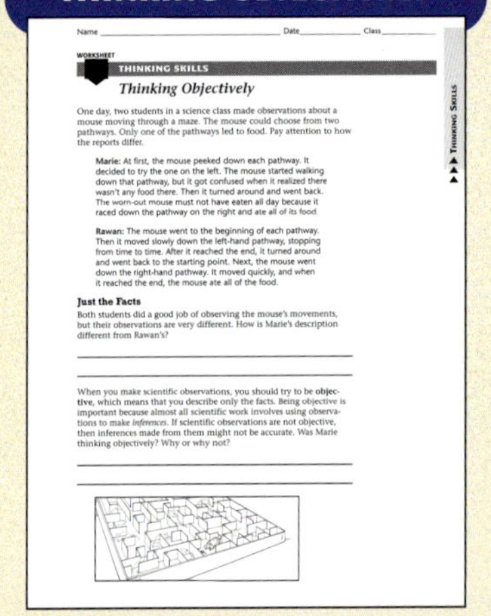
THINKING SKILLS

Thinking Objectively

UNDERSTANDING BIAS

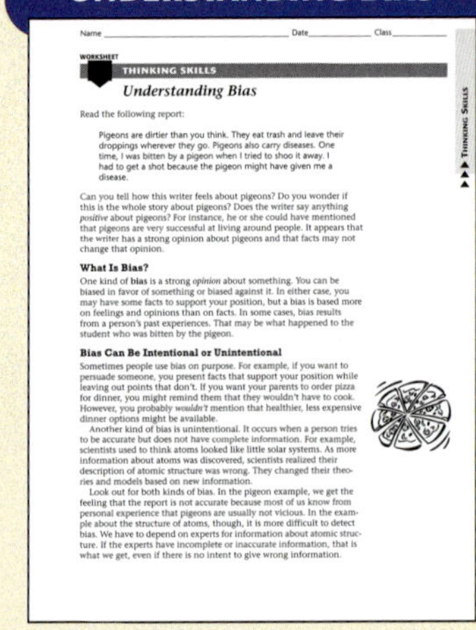
THINKING SKILLS

Understanding Bias

USING LOGIC

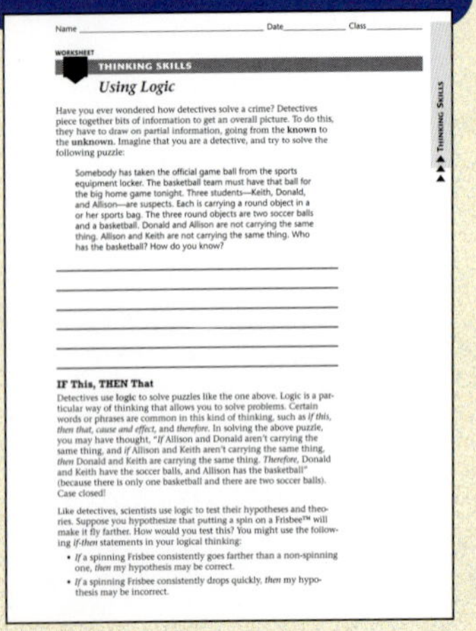
THINKING SKILLS

Using Logic

BOOSTING YOUR MEMORY

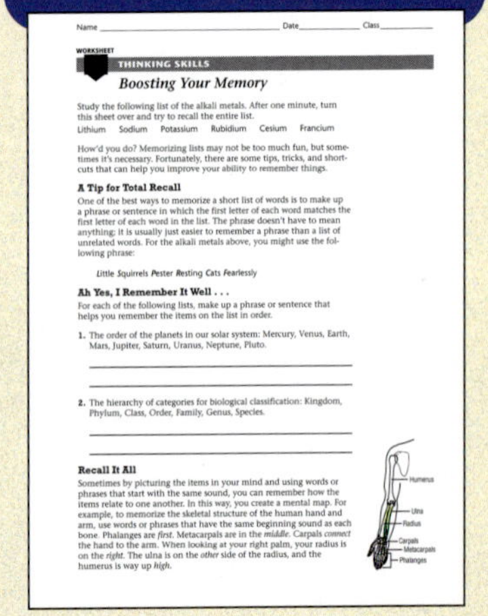
THINKING SKILLS

Boosting Your Memory

IMPROVING YOUR STUDY HABITS

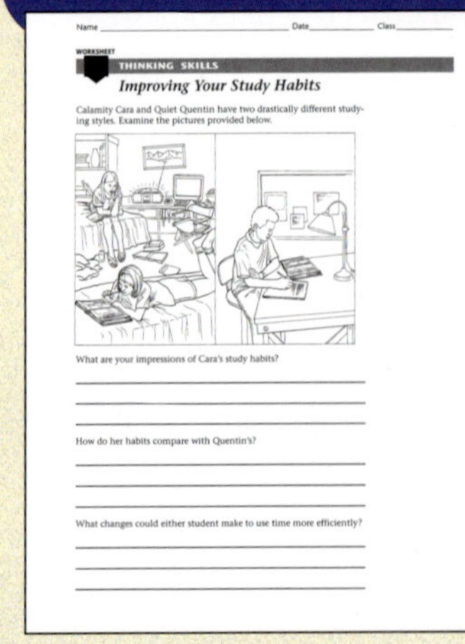
THINKING SKILLS

Improving Your Study Habits

READING A SCIENCE TEXTBOOK

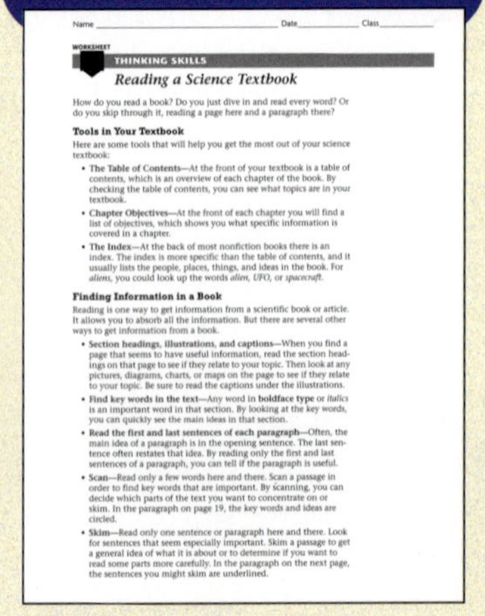
THINKING SKILLS

Reading a Science Textbook

Science Skills Worksheets: Experimenting Skills

SAFETY RULES!

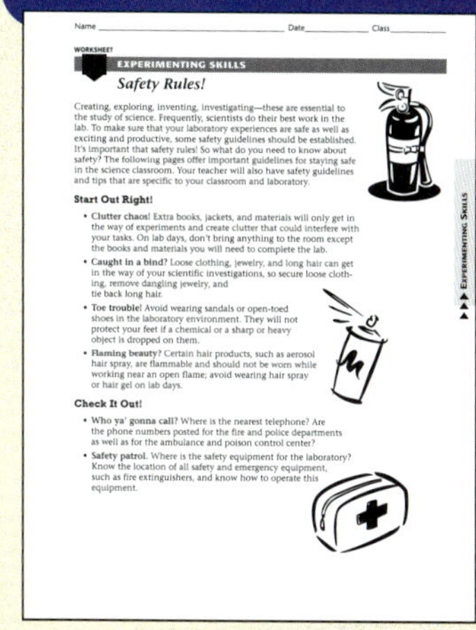
Safety Rules!

DOING A LAB WRITE-UP

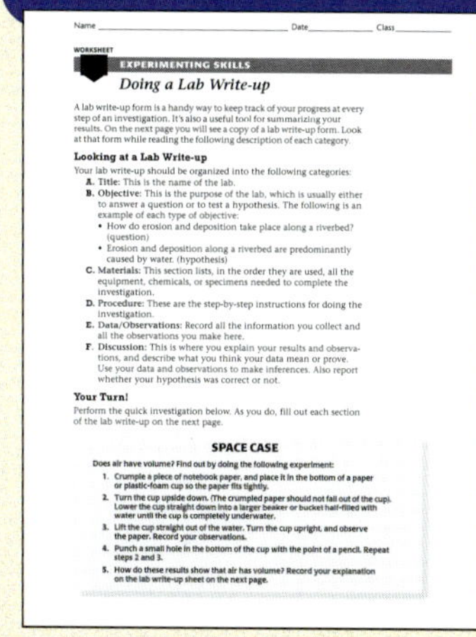
Doing a Lab Write-up

UNDERSTANDING VARIABLES

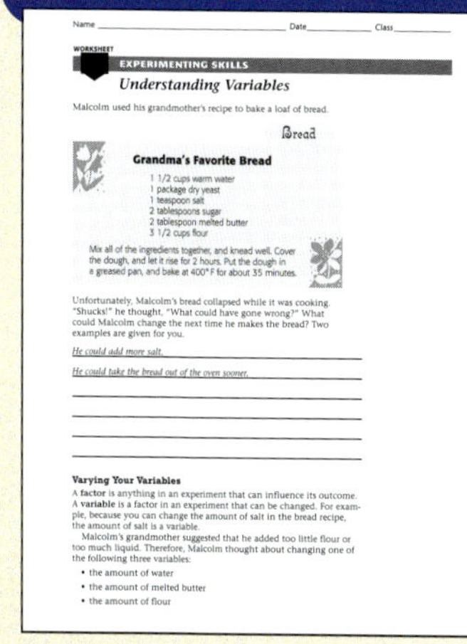
Understanding Variables

WORKING WITH HYPOTHESES

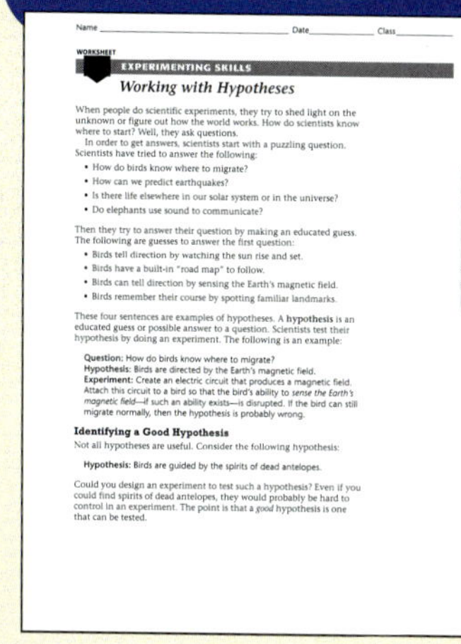
Working with Hypotheses

DESIGNING AN EXPERIMENT

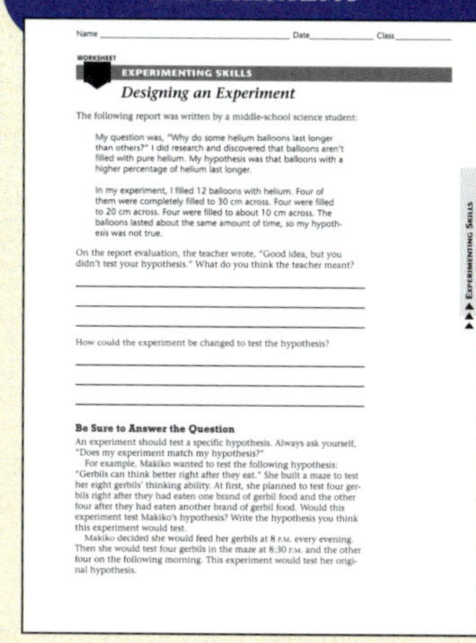
Designing an Experiment

USING THE INTERNATIONAL SYSTEM OF UNITS (SI)

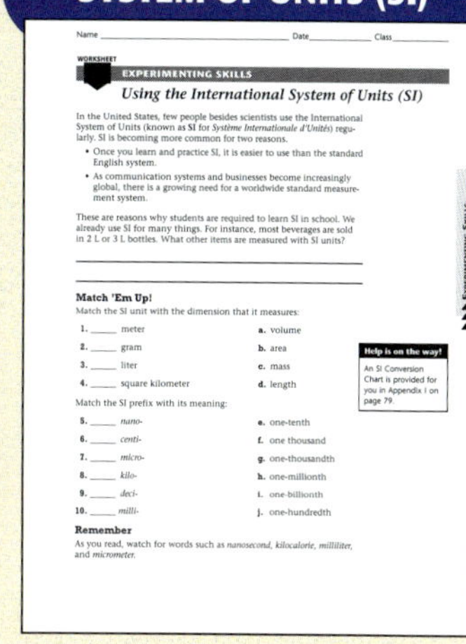
Using the International System of Units (SI)

MEASURING

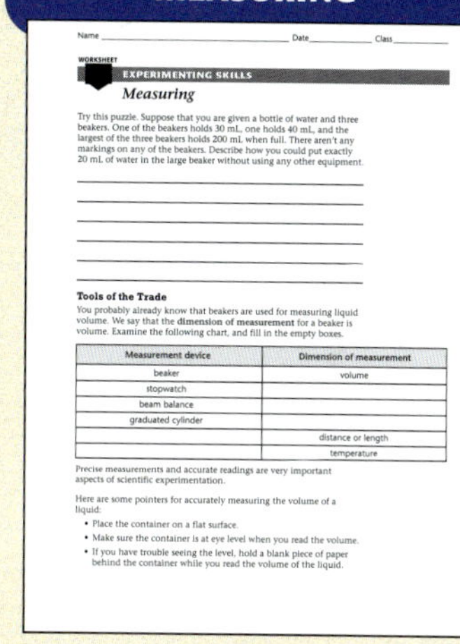
Measuring

Science Skills Worksheets: Researching Skills

CHOOSING YOUR TOPIC

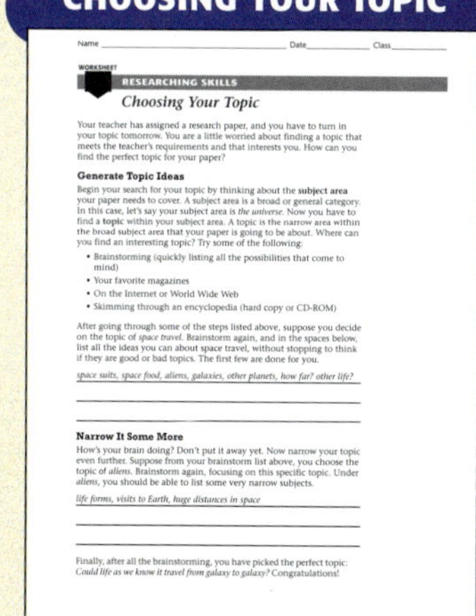
Choosing Your Topic

ORGANIZING YOUR RESEARCH

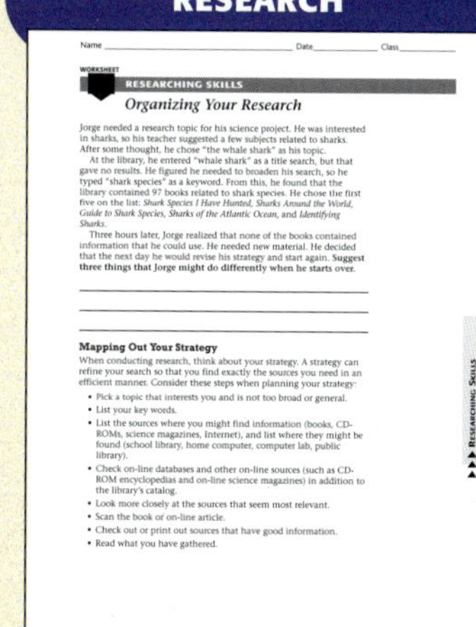
Organizing Your Research

FINDING USEFUL SOURCES

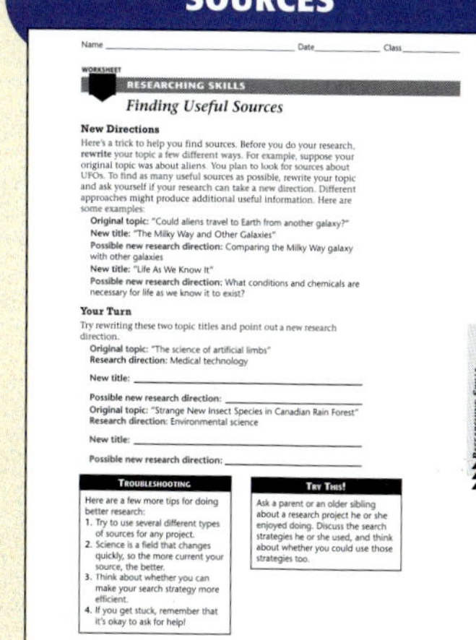
Finding Useful Sources

RESEARCHING ON THE WEB

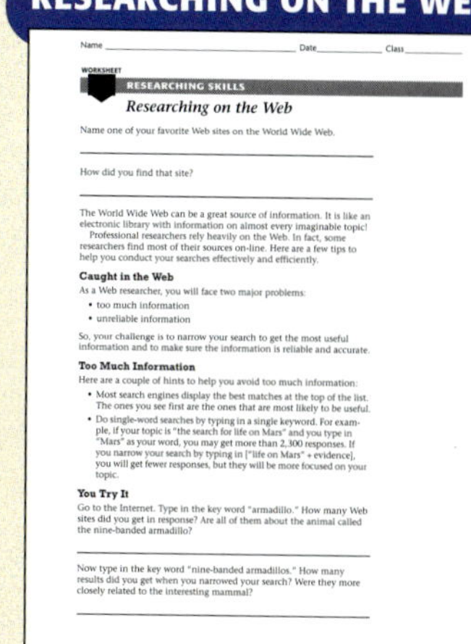
Researching on the Web

Science & Math Skills Worksheets (continued)

Science Skills Worksheets: Researching Skills (continued)

IDENTIFYING BIAS

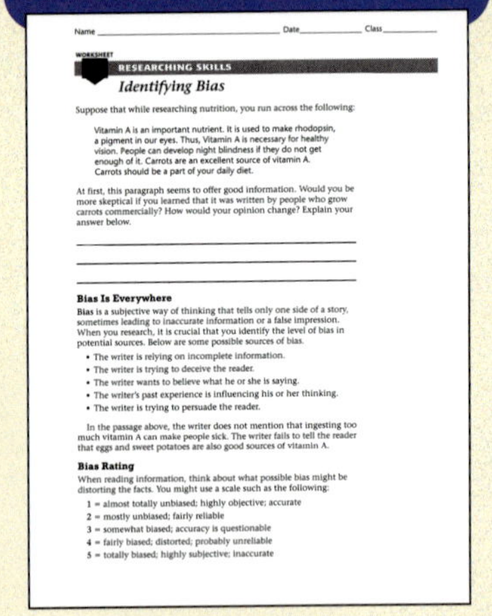

TAKING NOTES

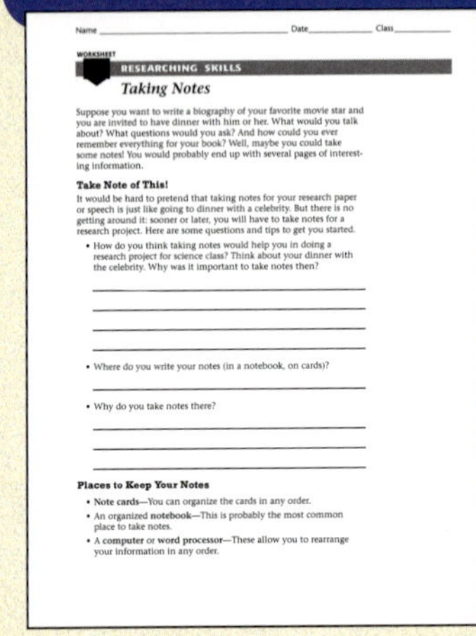

SCIENCE WRITING

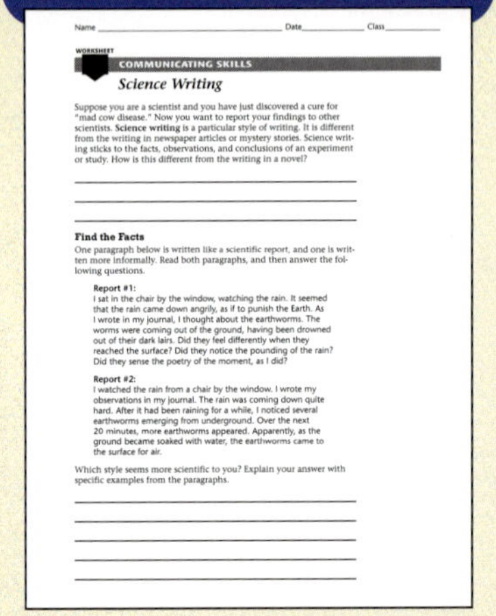

Science Skills Worksheets: Communicating Skills

SCIENCE DRAWING

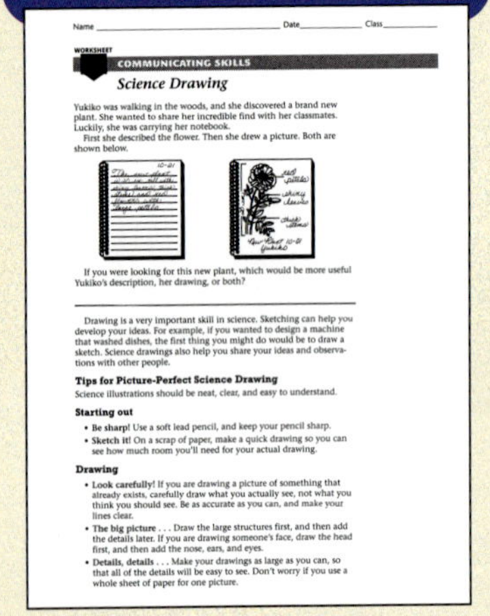

USING MODELS TO COMMUNICATE

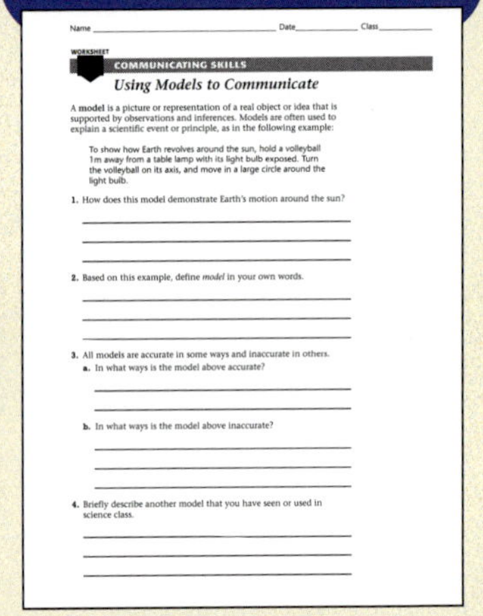

INTRODUCTION TO GRAPHS

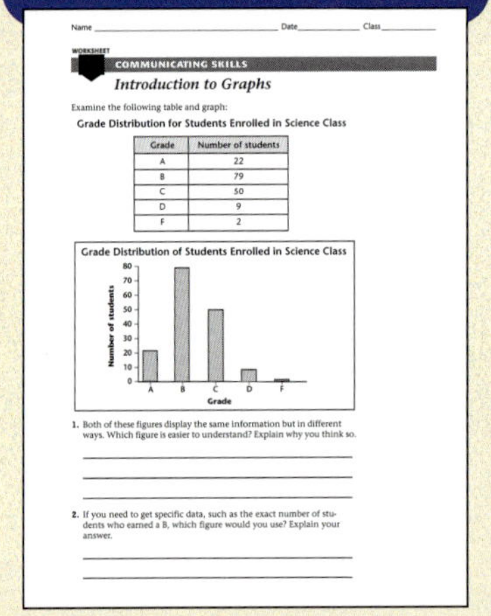

GRASPING GRAPHING

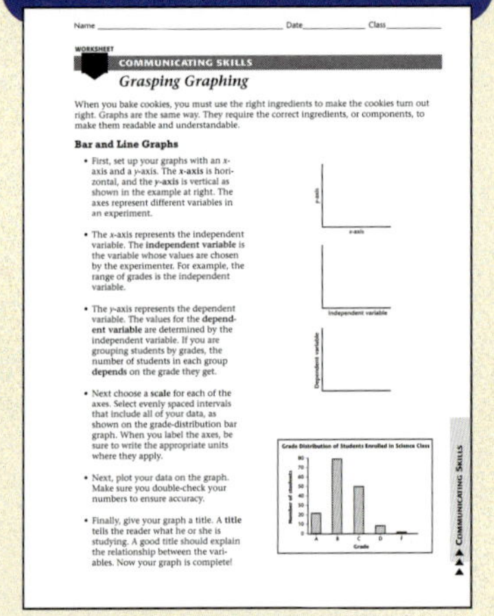

INTERPRETING YOUR DATA

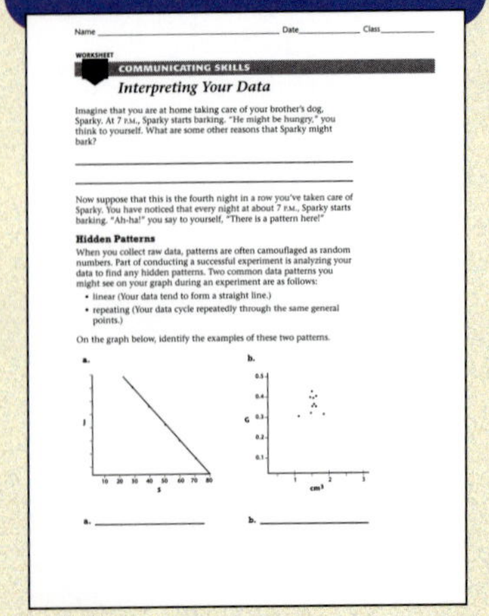

RECOGNIZING BIAS IN GRAPHS

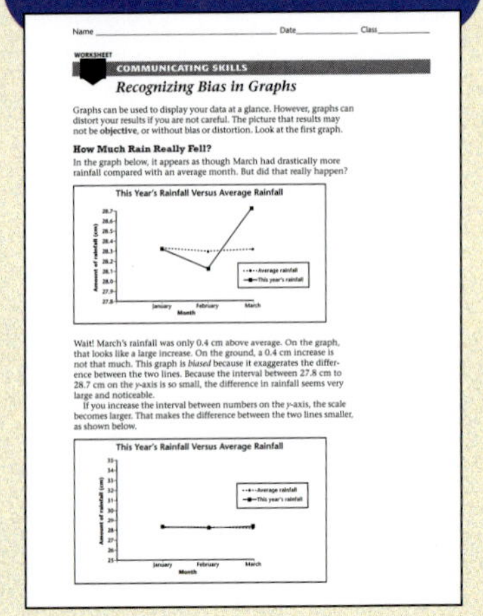

MAKING DATA MEANINGFUL

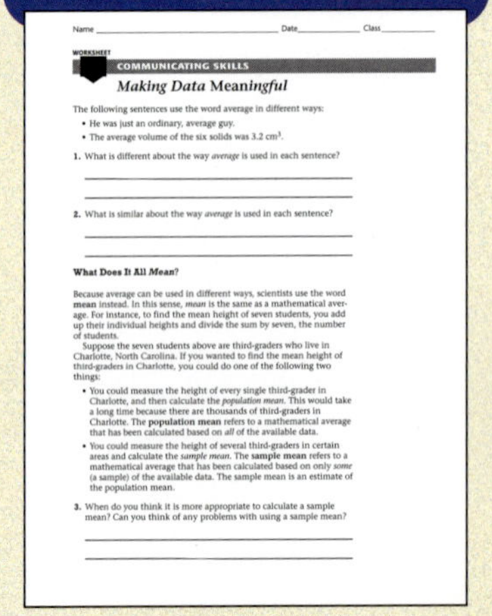

HINTS FOR ORAL PRESENTATIONS

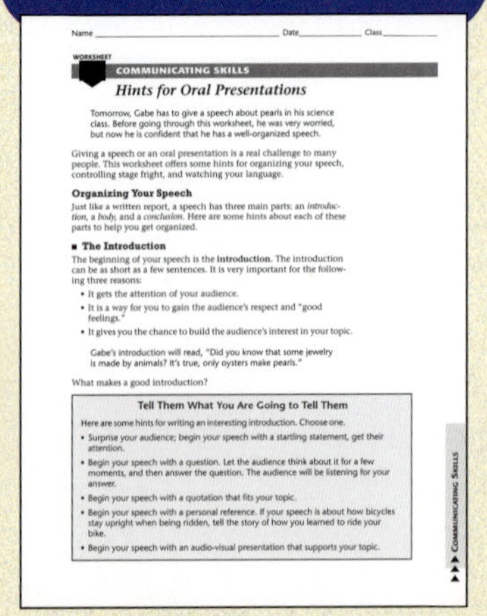

Math Skills for Science

ADDITION AND SUBTRACTION

Addition Review

Subtraction Review

MULTIPLICATION

Multiplying Whole Numbers

A Shortcut for Multiplying Large Numbers

DIVISION

Dividing Whole Numbers with Long Division

Checking Division with Multiplication

AVERAGES

What Is an Average?

Average, Mode, and Median

POSITIVE AND NEGATIVE NUMBERS

Comparing Integers on a Number Line

Arithmetic with Positive and Negative Numbers

FRACTIONS

What Is a Fraction?

Reducing Fractions to Lowest Terms

Improper Fractions and Mixed Numbers

Adding and Subtracting Fractions

Multiplying and Dividing Fractions

Science & Math Skills Worksheets (continued)

Math Skills for Science (continued)

RATIOS AND PROPORTIONS

MATH SKILLS — *What Is a Ratio?*

MATH SKILLS — *Using Proportions and Cross-Multiplication*

DECIMALS

MATH SKILLS — *Decimals and Fractions*

MATH SKILLS — *Arithmetic with Decimals*

PERCENTAGES

MATH SKILLS — *Parts of 100: Calculating Percentages*

MATH SKILLS — *Working with Percentages and Proportions*

MATH SKILLS — *Percentages, Fractions, and Decimals*

POWERS OF 10

MATH SKILLS — *Counting the Zeros*

MATH SKILLS — *Creating Exponents*

SCIENTIFIC NOTATION

MATH SKILLS — *What Is Scientific Notation?*

MATH SKILLS — *Multiplying and Dividing in Scientific Notation*

SI MEASUREMENT AND CONVERSION

MATH SKILLS — *What Is SI?*

MATH SKILLS — *A Formula for SI Catch-up*

Math Skills for Science (continued)

GEOMETRY

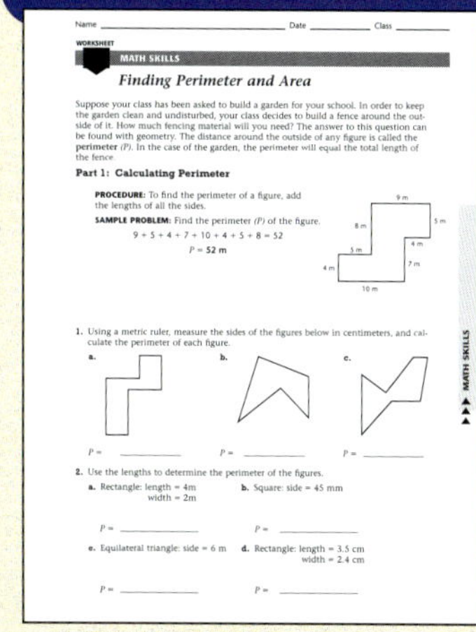

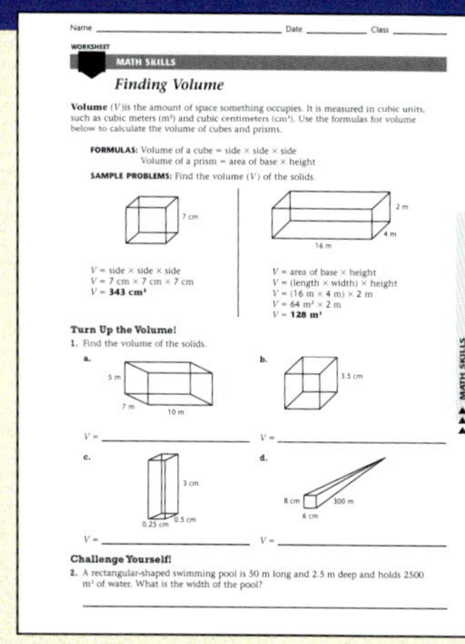

THE UNIT FACTOR AND DIMENSIONAL ANALYSIS

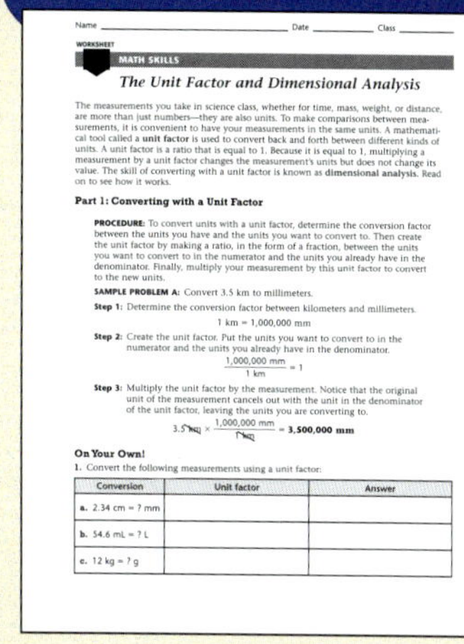

MATH IN SCIENCE: INTEGRATED SCIENCE

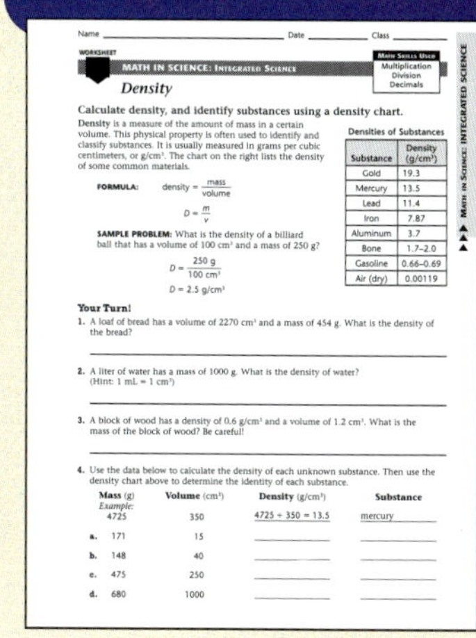

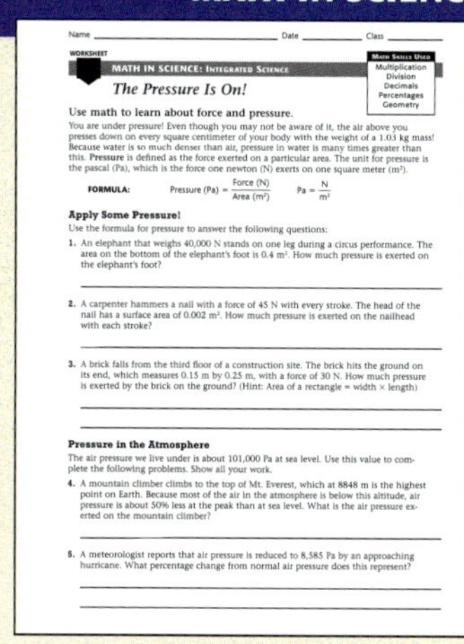

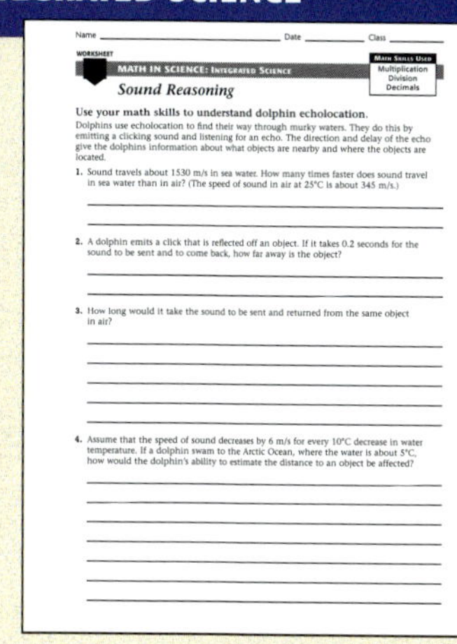

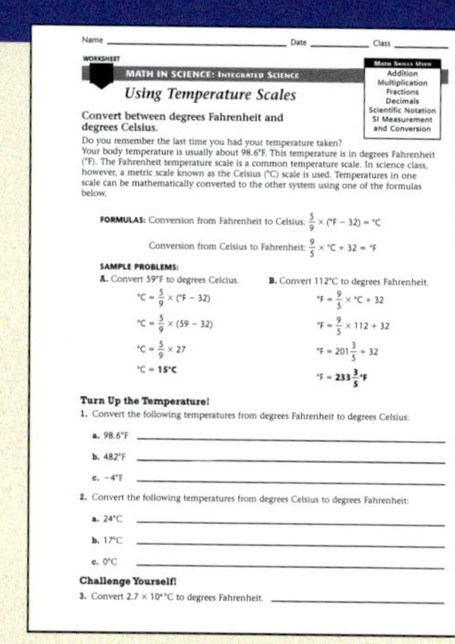

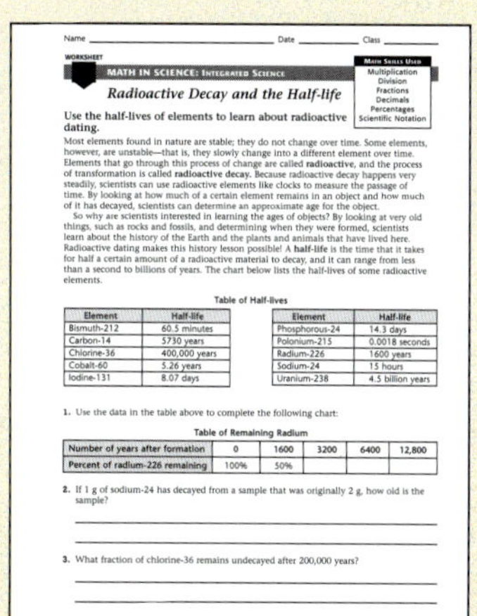

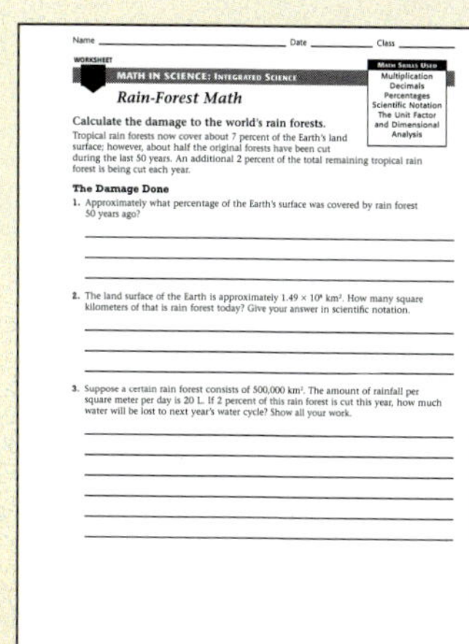

Science & Math Skills Worksheets (continued)

Math Skills for Science (continued)

MATH IN SCIENCE: EARTH SCIENCE

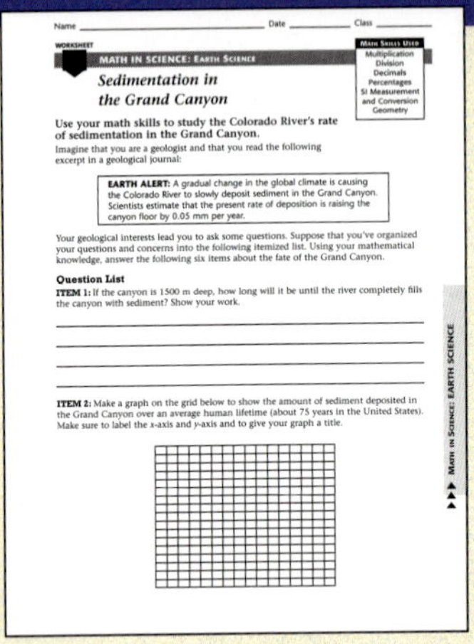
WORKSHEET — MATH IN SCIENCE: EARTH SCIENCE

Sedimentation in the Grand Canyon

Use your math skills to study the Colorado River's rate of sedimentation in the Grand Canyon.

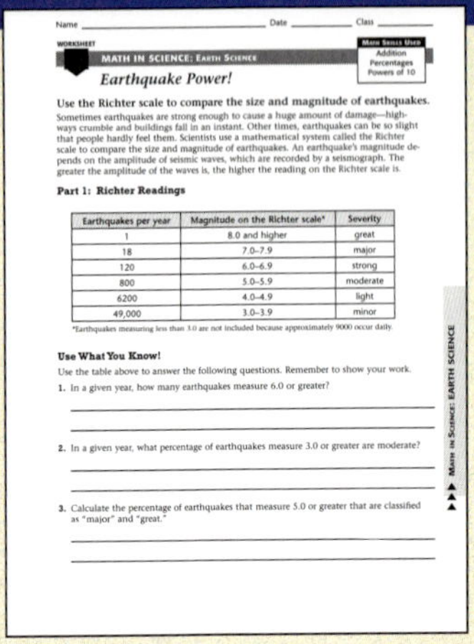
WORKSHEET — MATH IN SCIENCE: EARTH SCIENCE

Earthquake Power!

Use the Richter scale to compare the size and magnitude of earthquakes.

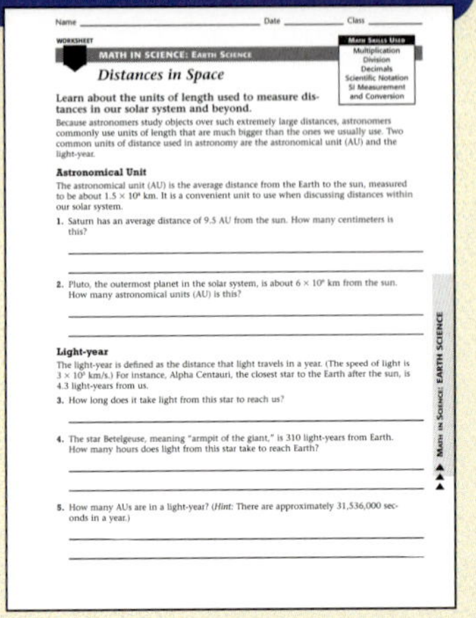
WORKSHEET — MATH IN SCIENCE: EARTH SCIENCE

Distances in Space

Learn about the units of length used to measure distances in our solar system and beyond.

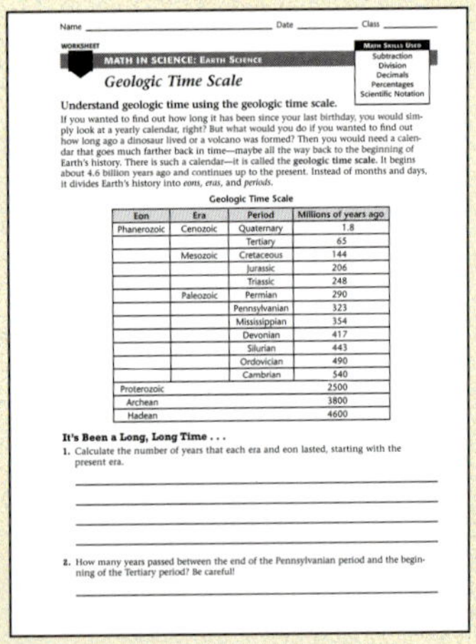
WORKSHEET — MATH IN SCIENCE: EARTH SCIENCE

Geologic Time Scale

Understand geologic time using the geologic time scale.

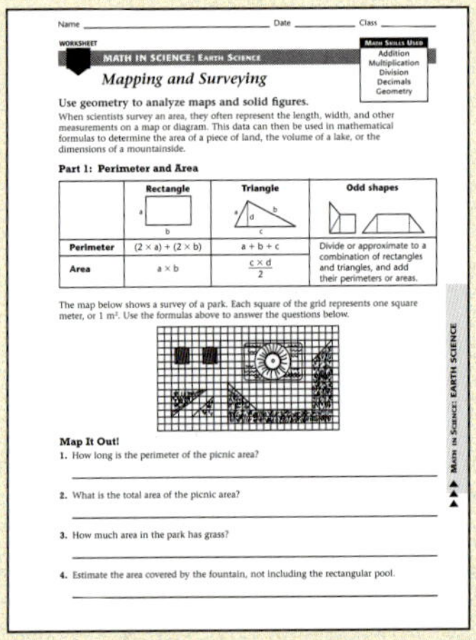
WORKSHEET — MATH IN SCIENCE: EARTH SCIENCE

Mapping and Surveying

Use geometry to analyze maps and solid figures.

Assessment Checklist & Rubrics

The following is just a sample of over 50 checklists and rubrics contained in this booklet.

RUBRICS FOR WRITTEN WORK

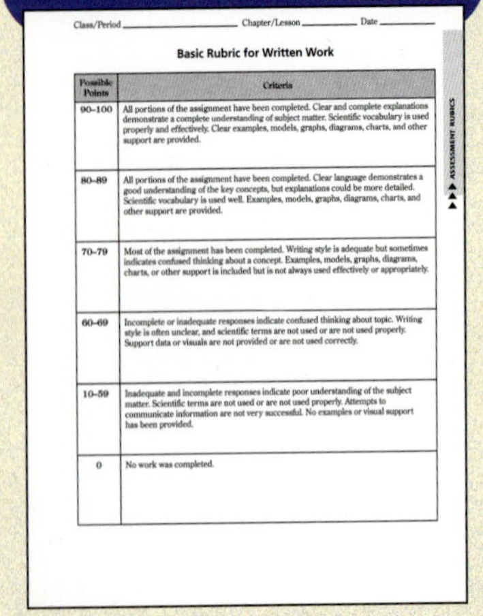
Basic Rubric for Written Work

RUBRIC FOR EXPERIMENTS

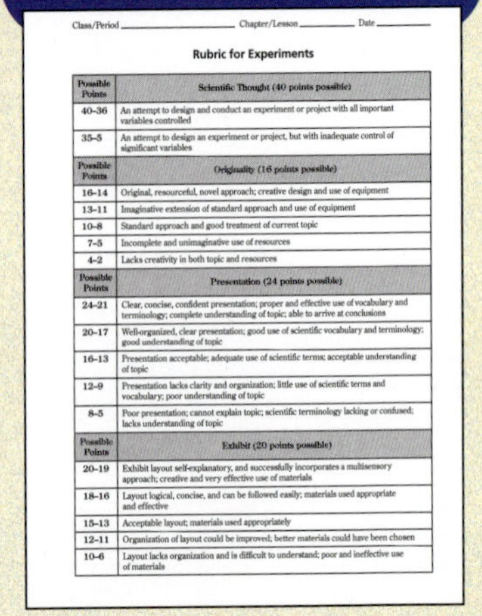
Rubric for Experiments

TEACHER EVALUATION OF COOPERATIVE LEARNING

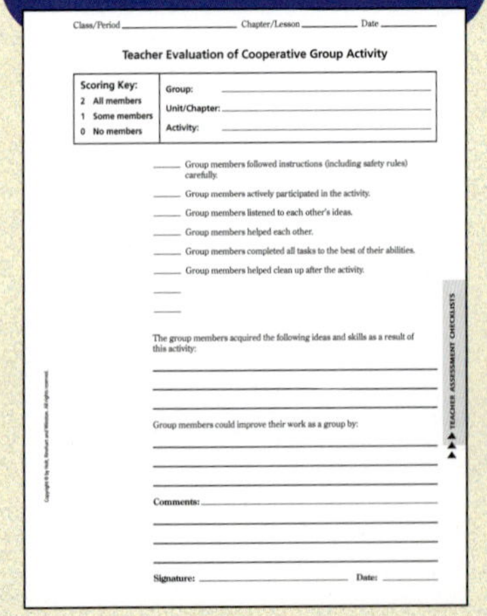
Teacher Evaluation of Cooperative Group Activity

TEACHER EVALUATION OF STUDENT PROGRESS

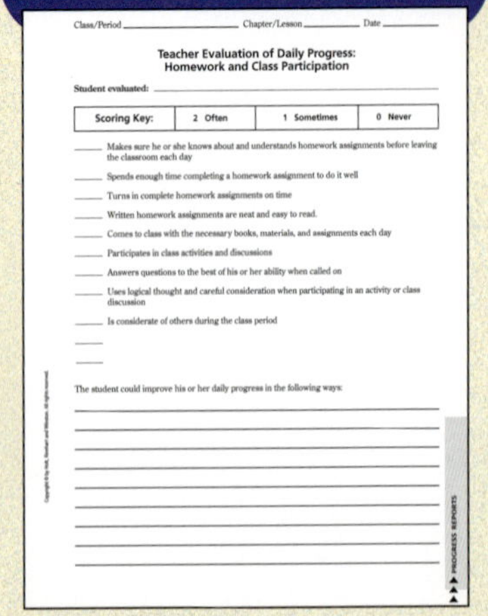
Teacher Evaluation of Daily Progress: Homework and Class Participation

HOLT SCIENCE & TECHNOLOGY

EARTH SCIENCE NATIONAL SCIENCE EDUCATION STANDARDS CORRELATIONS

The following lists show the chapter correlation of ***Holt Science and Technology: Water on Earth*** with the *National Science Education Standards* (grades 5-8)

UNIFYING CONCEPTS AND PROCESSES

Standard	Chapter Correlation
Systems, order, and organization Code: UCP 1	**Chapter 1** 1.1 **Chapter 3** 3.3
Evidence, models, and explanation Code: UCP 2	**Chapter 1** 1.1, 1.4 **Chapter 2** 2.2 **Chapter 3** 3.1, 3.3
Change, constancy, and measurement Code: UCP 3	**Chapter 1** 1.4 **Chapter 2** 2.2 **Chapter 3** 3.2
Form and function Code: UCP 5	**Chapter 1** 1.1

SCIENCE AS INQUIRY

Standard	Chapter Correlation
Abilities necessary to do scientific inquiry Code: SAI 1	**Chapter 1** 1.1, 1.3, 1.4 **Chapter 2** 2.2, 2.4 **Chapter 3** 3.1, 3.2, 3.3

HISTORY AND NATURE OF SCIENCE

Standard	Chapter Correlation
Science as a human endeavor Code: HNS 1	**Chapter 2** 2.2 **Chapter 3** 3.1, 3.3
History of science Code: HNS 3	**Chapter 2** 2.2 **Chapter 3** 3.1, 3.3

SCIENCE IN PERSONAL AND SOCIAL PERSPECTIVES

Standard	Chapter Correlation
Personal health Code: SPSP 1	**Chapter 1** 1.4
Populations, resources, and environments Code: SPSP 2	**Chapter 1** 1.3, 1.4 **Chapter 2** 2.4, 2.5
Natural hazards Code: SPSP 3	**Chapter 1** 1.2 **Chapter 3** 3.2
Risks and benefits Code: SPSP 4	**Chapter 1** 1.3 **Chapter 2** 2.4, 2.5 **Chapter 3** 3.2
Science and technology in society Code: SPSP 5	**Chapter 1** 1.4 **Chapter 2** 2.4, 2.5

SCIENCE AND TECHNOLOGY

Standard	Chapter Correlation
Abilities of technological design Code: ST 1	**Chapter 2** 2.2 **Chapter 3** 3.3
Understandings about science and technology Code: ST 2	**Chapter 1** 1.4 **Chapter 2** 2.2, 2.4, 2.5

EARTH SCIENCE NATIONAL SCIENCE EDUCATION CONTENT STANDARDS

STRUCTURE OF THE EARTH SYSTEM

Standard	Chapter Correlation
Lithospheric plates on the scales of continents and oceans constantly move at rates of centimeters per year in response to movements in the mantle. Major geological events, such as earthquakes, volcanic eruptions, and mountain building result from these plate motions. Code: ES 1b	**Chapter 2** 2.1, 2.2 **Chapter 3** 3.2
Land forms are the result of a combination of constructive and destructive forces. Constructive forces include crustal deformation, volcanic eruption, and deposition of sediment, while destructive forces include weathering and erosion. Code: ES 1c	**Chapter 1** 1.1, 1.2, 1.3, 1.4 **Chapter 2** 2.2
Some changes in the solid earth can be described as the "rock cycle." Old rocks at the earth's surface weather, forming sediments that are buried, then compacted, heated, and often recrystallized into new rock. Eventually, those new rocks may be brought to the earth's surface by the forces that drive plate motions, and the rock cycle continues. Code: ES 1d	**Chapter 2** 2.1
Water, which covers the majority of the earth's surface, circulates through the crust, oceans, and atmosphere in what is known as the "water cycle." Water evaporates from the earth's surface, rises and cools as it moves to higher elevations, condenses as rain or snow, and falls to the surface where it collects in lakes, oceans, soil, and in rocks underground. Code: ES 1f	**Chapter 1** 1.1, 1.3 **Chapter 2** 2.1
Water is a solvent. As it passes through the water cycle it dissolves minerals and gases and carries them to the oceans. Code: ES 1g	**Chapter 1** 1.1, 1.3 **Chapter 2** 2.1
The atmosphere is a mixture of nitrogen, oxygen, and trace gases that include water vapor. The atmosphere has different properties at different elevations. Code: ES 1h	**Chapter 2** 2.1
Global patterns of atmospheric movement influence local weather. Oceans have a major effect on climate, because water in the oceans holds a large amount of heat. Code: ES 1j	**Chapter 2** 2.1 **Chapter 3** 3.1

EARTH'S HISTORY

Standard	Chapter Correlation
The earth processes we see today, including erosion, movement of lithospheric plates, and changes in atmospheric composition, are similar to those that occurred in the past. Earth history is also influenced by occasional catastrophes, such as the impact of an asteroid or comet. Code: ES 2a	**Chapter 2** 2.1

EARTH IN THE SOLAR SYSTEM

Standard	Chapter Correlation
Gravity is the force that keeps planets in orbit around the sun and governs the rest of the motion in the solar system. Gravity alone holds us to the earth's surface and explains the phenomenon of the tides. Code: ES 3c	**Chapter 3** 3.3
The sun is the major source of energy for phenomena on the earth's surface, such as growth of plants, winds, ocean currents, and the water cycle. Seasons result from variations in the amount of the sun's energy hitting the surface, due to the tilt of the earth's rotation on its axis and the length of the day. Code: ES 3d	**Chapter 3** 3.1

Master Materials List

For added convenience, Science Kit® provides materials-ordering software on CD-ROM designed specifically for *Holt Science and Technology*. Using this software, you can order complete kits or individual items, quickly and efficiently.

Consumable Materials	Amount	Page
Bag, plastic sealable sandwich	1	100
Bottle, soda, 2 L, with cap	1	100
Cardboard, 1 x 1 cm	1	106
Cardboard, corrugated, disk 5 cm diam.	1	106
Cardboard, corrugated, disk 15 cm diam.	1	106
Charcoal, activated	1/2 lb	100
Cup, clear plastic, 9 fl oz	1	33
Cup, paper	1	3
Cup, plastic-foam	5	15
Detergent, dishwashing (powder or liquid)	40 mL	100
Detritus (cut up leaves and grass)	40 mL	100
Dowel, wood, 1/4 in. diam., 36 cm	1	106
Food coloring, blue	1 bottle	69, 88
Food coloring, red	1 bottle	69, 88
Gloves, protective	1	104
Glue, white	1 bottle	106
Gravel	2 lb	3
Ice	1 bucket	88, 100
Oil, light machine	15 mL	104
pH test strips	6	100
Plastic wrap, 30 x 20 cm	4	88
Rope, approx. 3 m long	1	82
Salt	approx. 1 tbsp	88
Sand	4 oz	15
Sand	1/2 lb	100
Sand	15 lb	3
Soil, clay	40 mL	100
Soil, potting	4 oz	15
String, 5 cm	1	106
Vinegar	40 mL	100

Nonconsumable Equipment	Amount	Page
Beaker, 100 mL	1	24
Beaker, 400 mL	1	100
Beaker, 400 mL	5	88
Beaker, 1000 mL	1	53
Beaker, 1000 mL	2	100
Bowl, clear plastic, 3 gal	1	33
Calculator, scientific	1	104
Clay, modeling	2 lb	60
Clothespin	1	3
Dropper pipet	2	104
Erlenmeyer flask	1	100
Gloves, heat-resistant	2	24, 88, 100
Graduated cylinder, 50 mL	1	24, 53, 104
Graduated cylinder, 100 mL	1	53
Graduated cylinder, 1000 mL	1	53
Gravel	1/2 lb	100
Hammer	1	100
Hot plate	1	24, 88, 100
Magnifying lens	1	3, 100
Nail, approx. 1 in.	1	100
Pan, aluminum pie, 22 cm diam.	1	104
Pencil, unsharpened	8	60
Plastic tubing, 5 mm diam., 1.5 m long	1	100
Ribbon, approx. 30 cm long	1	82
Ring stand with ring	1	100
Ruler, metric	1	60, 104
Scissors	1	60, 100
Shoe box with lid	1	60
Spoon, plastic	2	88, 100
Stapler, with staples	1	106
Stopper, rubber, 1 hole, with glass tube	1	100
Stopwatch	1	88
Tongs	1	24
Washtub, plastic	1	3, 69
Watch glass	1	24

Answers to Concept Mapping Questions

The following pages contain sample answers to all of the concept mapping questions that appear in the Chapter Reviews. Because there is more than one way to do a concept map, your students' answers may vary.

CHAPTER 1 The Flow of Fresh Water

19.

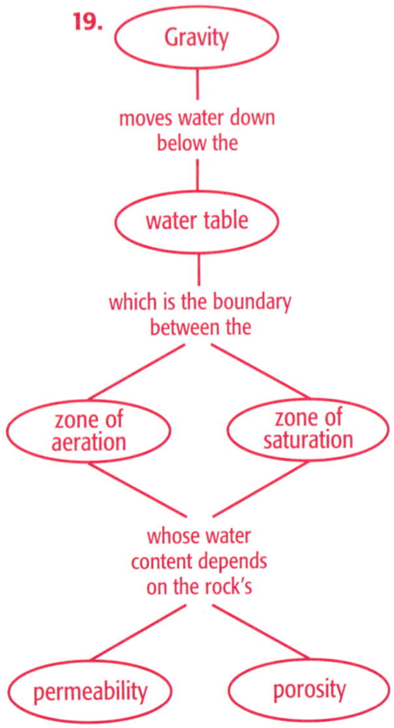

CHAPTER 2 Exploring the Oceans

19.

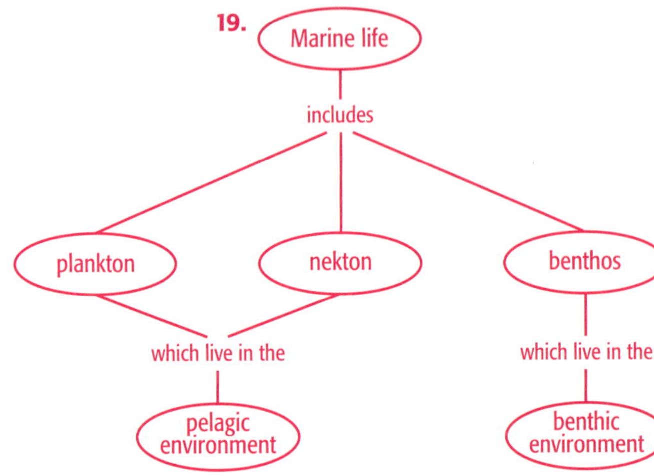

CHAPTER 3 The Movement of Ocean Water

19.

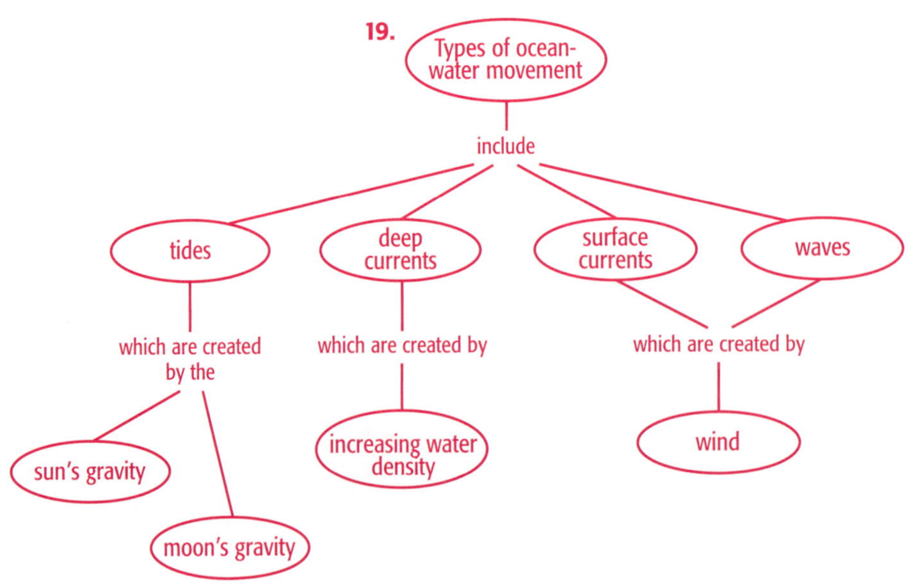

To the Student

This book was created to make your science experience interesting, exciting, and fun!

Go for It!

Science is a process of discovery, a trek into the unknown. The skills you develop using *Holt Science & Technology*—such as observing, experimenting, and explaining observations and ideas—are the skills you will need for the future. There is a universe of exploration and discovery awaiting those who accept the challenges of science.

Science & Technology

You see the interaction between science and technology every day. Science makes technology possible. On the other hand, some of the products of technology, such as computers, are used to make further scientific discoveries. In fact, much of the scientific work that is done today has become so technically complicated and expensive that no one person can do it entirely alone. But make no mistake, the creative ideas for even the most highly technical and expensive scientific work still come from individuals.

Activities and Labs

The activities and labs in this book will allow you to make some basic but important scientific discoveries on your own. You can even do some exploring on your own at home! Here's your chance to use your imagination and curiosity as you investigate your world.

Keep a ScienceLog

In this book, you will be asked to keep a type of journal called a ScienceLog to record your thoughts, observations, experiments, and conclusions. As you develop your ScienceLog, you will see your own ideas taking shape over time. You'll have a written record of how your ideas have changed as you learn about and explore interesting topics in science.

Know "What You'll Do"

The "What You'll Do" list at the beginning of each section is your built-in guide to what you need to learn in each chapter. When you can answer the questions in the Section Review and Chapter Review, you know you are ready for a test.

Check Out the Internet

You will see this sciLINKS logo throughout the book. You'll be using sciLINKS as your gateway to the Internet. Once you log on to sciLINKS using your computer's Internet link, type in the sciLINKS address. When asked for the keyword code, type in the keyword for that topic. A wealth of resources is now at your disposal to help you learn more about that topic.

In addition to sciLINKS you can log on to some other great resources to go with your text. The addresses shown below will take you to the home page of each site.

internet connect

This textbook contains the following on-line resources to help you make the most of your science experience.

go.hrw.com	sciLINKS NSTA	Smithsonian Institution® Internet Connections	CNNfyi.com
Visit **go.hrw.com** for extra help and study aids matched to your textbook. Just type in the keyword HR2 HOME.	Visit **www.scilinks.org** to find resources specific to topics in your textbook. Keywords appear throughout your book to take you further.	Visit **www.si.edu/hrw** for specifically chosen on-line materials from one of our nation's premier science museums.	Visit **www.cnnfyi.com** for late-breaking news and current events stories selected just for you.

Chapter Organizer

CHAPTER ORGANIZATION	TIME MINUTES	OBJECTIVES	LABS, INVESTIGATIONS, AND DEMONSTRATIONS
Chapter Opener **pp. 2–3**	45	National Standards: SAI 1	**Start-Up Activity,** Stream Weavers, p. 3
Section 1 **The Active River**	90	▶ Illustrate the water cycle. ▶ Describe a drainage basin. ▶ Explain the major factors that affect the rate of stream erosion. ▶ Identify the stages of river development. UCP 1, 5, SAI 1, ES 1c, 1f, 1g, 2a; Labs UCP 2, SAI 1, ES 1f	**Demonstration,** p. 5 in ATE **Making Models,** Water Cycle—What Goes Up . . . , p. 24 **Datasheets for LabBook,** Water Cycle—What Goes Up . . .
Section 2 **Stream and River Deposits**	90	▶ Describe the different types of stream deposits. ▶ Explain the relationship between rich agricultural regions and river flood plains. SPSP 3, 4, ES 1c	**Demonstration,** Modeling Deposition, p. 11 in ATE
Section 3 **Water Underground**	90	▶ Identify and describe the location of a water table. ▶ Describe the characteristics of an aquifer. ▶ Explain how caves and sinkholes form as a result of erosion and deposition. SAI 1, SPSP 2, ES 1c, 1f, 1g	**Demonstration,** p. 14 in ATE **QuickLab,** Degree of Permeability, p. 15
Section 4 **Using Water Wisely**	90	▶ Describe the stages of treatment for water at a sewage treatment plant. ▶ Compare a septic system with a sewage treatment plant. ▶ Explain how ground water can be both a renewable and a nonrenewable resource. SAI 1, SPSP 1, 2, 5, ES 1c; Labs UCP 2, 3, ST 2, SAI 1, SPSP 5	**Interactive Explorations CD-ROM,** Flood Bank *A **Worksheet** is also available in the **Interactive Explorations Teacher's Edition.*** **Discovery Lab,** Clean Up Your Act, p. 100 **Datasheets for LabBook,** Clean Up Your Act **EcoLabs & Field Activities,** The Frogs Are Off Course **Long-Term Projects & Research Ideas,** Canyon Controversy

*See page **T23** for a complete correlation of this book with the*

NATIONAL SCIENCE EDUCATION STANDARDS.

TECHNOLOGY RESOURCES

Guided Reading Audio CD
English or Spanish, Chapter 1

One-Stop Planner CD-ROM with Test Generator

Interactive Explorations CD-ROM
CD 1, Exploration 8, Flood Bank

Earth Science Videodisc
Water and Erosion: 46567–48921

CNN. **Science, Technology & Society,** Tapping into Yellowstone's Hot Springs, Segment 1
Eye on the Environment, Watch for Flooding, Segment 9
China's Superdam, Segment 10

Science Discovery Videodiscs
Image and Activity Bank with Lesson Plans: Dirty Water
Science Sleuths: The Missing Beach

Chapter 1 • The Flow of Fresh Water

CLASSROOM WORKSHEETS, TRANSPARENCIES, AND RESOURCES	SCIENCE INTEGRATION AND CONNECTIONS	REVIEW AND ASSESSMENT
Directed Reading Worksheet **Science Puzzlers, Twisters & Teasers**		
Math Skills for Science Worksheet, Checking Division with Multiplication **Transparency 141,** The Water Cycle **Directed Reading Worksheet,** Section 1 **Transparency 75,** River Features	**Real-World Connection,** p. 6 in ATE **MathBreak,** Calculating a Stream's Gradient, p. 7 **Connect to Physical Science,** p. 7 in ATE **Cross-Disciplinary Focus,** p. 9 in ATE	**Self-Check,** p. 8 **Homework,** p. 8 in ATE **Section Review,** p. 10 **Quiz,** p. 10 in ATE **Alternative Assessment,** p. 10 in ATE
Directed Reading Worksheet, Section 2 **Reinforcement Worksheet,** Fresh Water in the United States	**Astronomy Connection,** p. 12	**Self-Check,** p. 12 **Homework,** p. 12 in ATE **Section Review,** p. 13 **Quiz,** p. 13 in ATE **Alternative Assessment,** p. 13 in ATE
Transparency 142, The Water Table **Directed Reading Worksheet,** Section 3 **Transparency 143,** Artesian Formation **Transparency 143,** The Water Table and Wells **Reinforcement Worksheet,** Dig It! **Problem Solving Worksheet,** Water Crisis at Happy Acres	**Multicultural Connection,** p. 14 in ATE **Real-World Connection,** p. 15 in ATE **Connect to Life Science,** p. 16 in ATE **Environment Connection,** p. 17 **Weird Science:** Bubble, Boil, & Squirt, p. 30	**Self-Check,** p. 16 **Section Review,** p. 18 **Quiz,** p. 18 in ATE **Alternative Assessment,** p. 18 in ATE
Directed Reading Worksheet, Section 4 **Math Skills for Science Worksheet,** Multiplying Whole Numbers	**Multicultural Connection,** p. 20 in ATE **Math and More,** p. 21 in ATE **Connect to Environmental Science** p. 22 in ATE **Apply,** p. 22 **Eye on the Environment:** Disaster Along the Delta, p. 31	**Homework,** p. 20 in ATE **Section Review,** p. 23 **Quiz,** p. 23 in ATE **Alternative Assessment,** p. 23 in ATE

internetconnect

go.hrw.com — **Holt, Rinehart and Winston On-line Resources**

go.hrw.com

For worksheets and other teaching aids related to this chapter, visit the HRW Web site and type in the keyword: **HSTDEP**

sciLINKS NSTA — **National Science Teachers Association**

www.scilinks.org

Encourage students to use the *sci*LINKS numbers listed in the internet connect boxes to access information and resources on the **NSTA** Web site.

END-OF-CHAPTER REVIEW AND ASSESSMENT

Chapter Review in Study Guide
Vocabulary and Notes in Study Guide
Chapter Tests with Performance-Based Assessment, Chapter 1 Test
Chapter Tests with Performance-Based Assessment, Performance-Based Assessment 1
Concept Mapping Transparency 11

Chapter Resources & Worksheets

Visual Resources

TEACHING TRANSPARENCIES

#141

#142

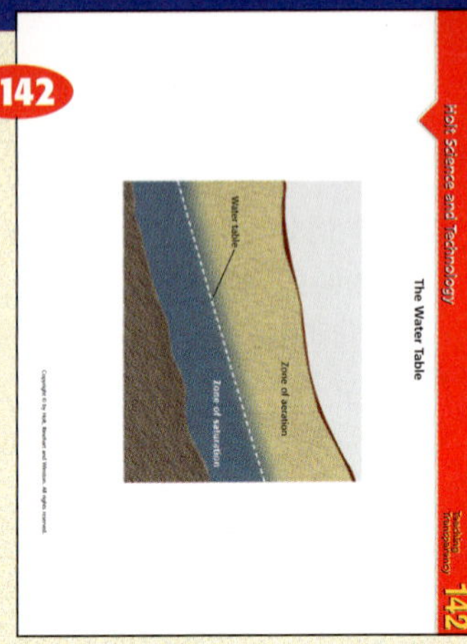

#143

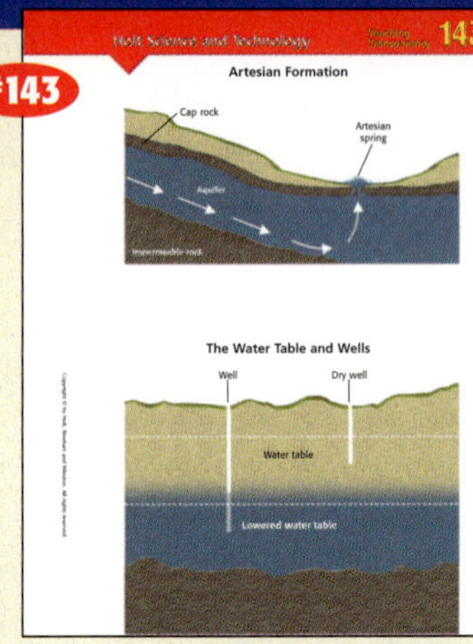

TEACHING TRANSPARENCIES

#75

CONCEPT MAPPING TRANSPARENCY

#11

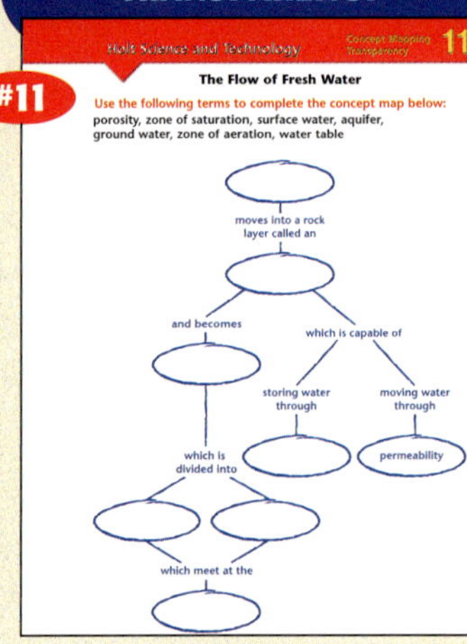

Meeting Individual Needs

DIRECTED READING

#1

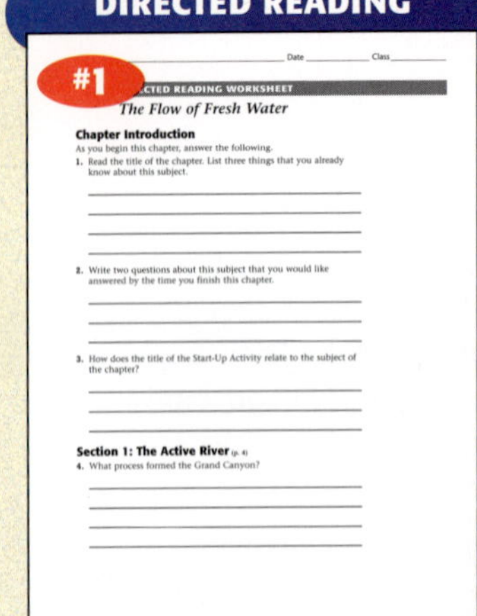

REINFORCEMENT & VOCABULARY REVIEW

#1

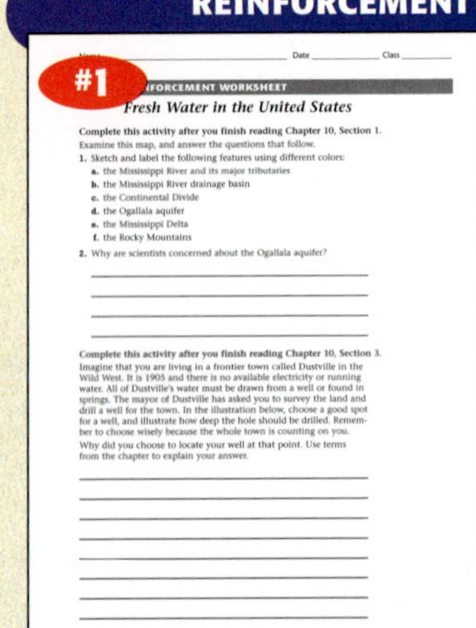

#1

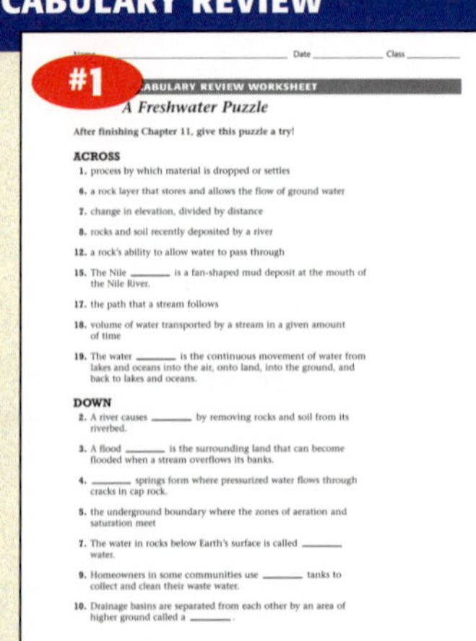

SCIENCE PUZZLERS, TWISTERS & TEASERS

#1

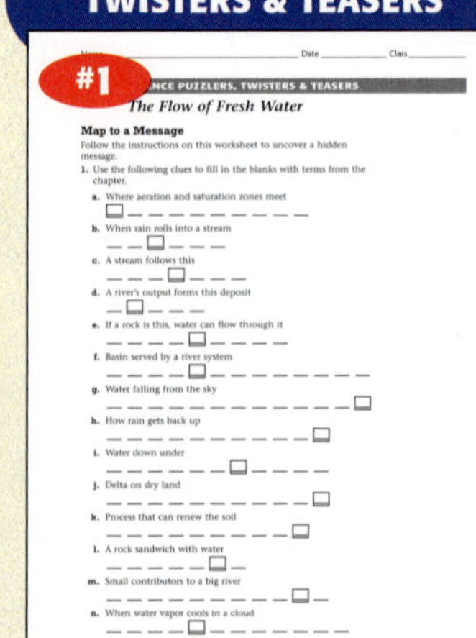

Chapter 1 • The Flow of Fresh Water

Review & Assessment

STUDY GUIDE

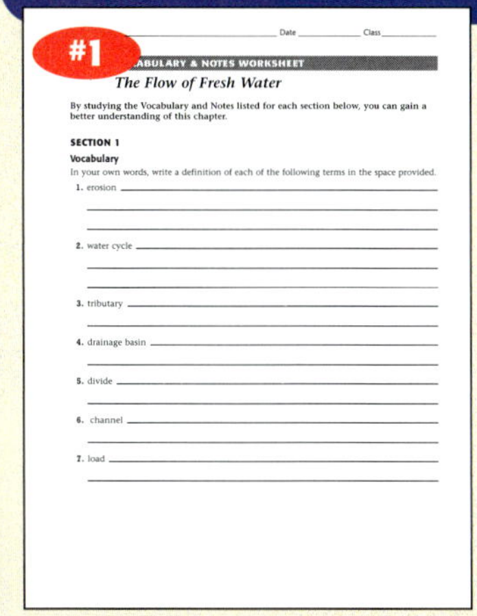
#1
VOCABULARY & NOTES WORKSHEET
The Flow of Fresh Water

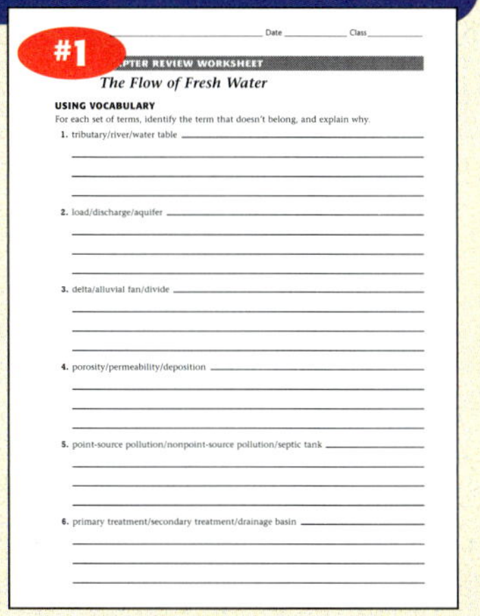
#1
CHAPTER REVIEW WORKSHEET
The Flow of Fresh Water

CHAPTER TESTS WITH PERFORMANCE-BASED ASSESSMENT

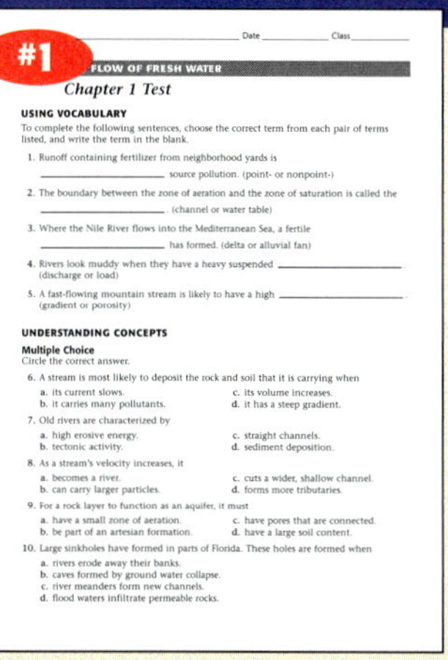
#1
THE FLOW OF FRESH WATER
Chapter 1 Test

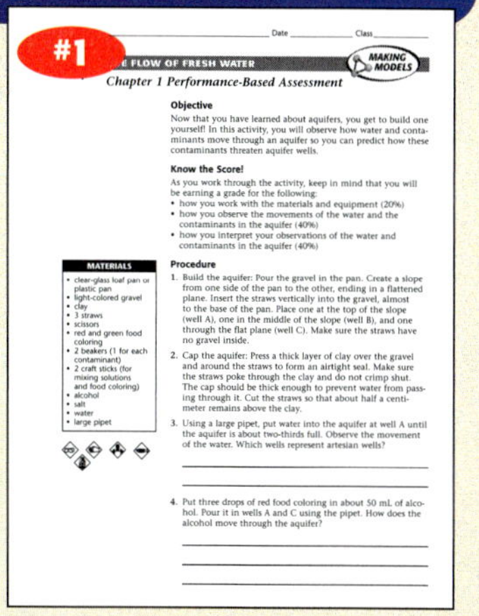
#1
THE FLOW OF FRESH WATER
Chapter 1 Performance-Based Assessment

Lab Worksheets

ECOLABS & FIELD ACTIVITIES

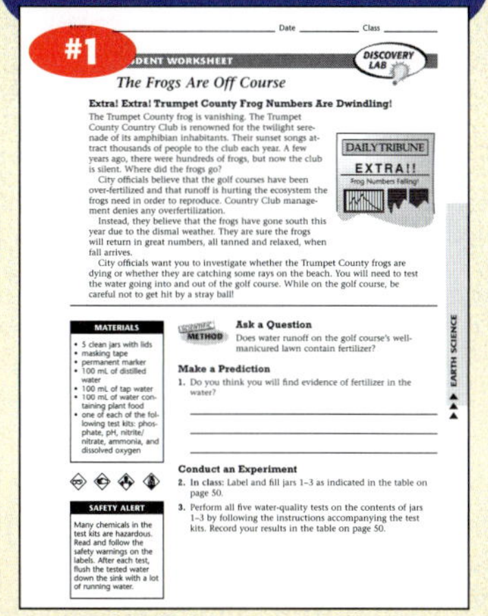
#1
The Frogs Are Off Course

LONG-TERM PROJECTS & RESEARCH IDEAS

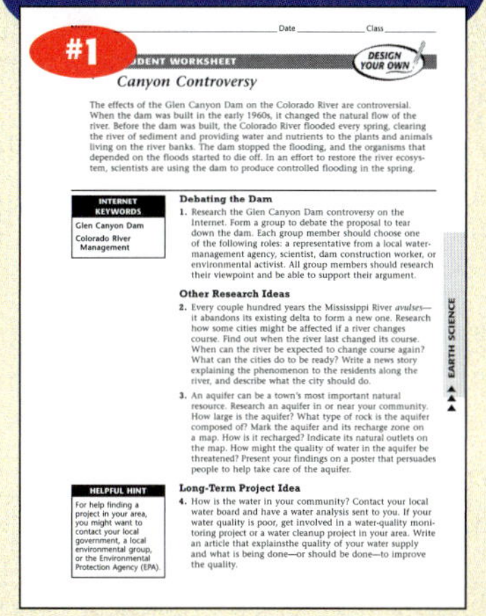
#1
Canyon Controversy

DATASHEETS FOR LABBOOK

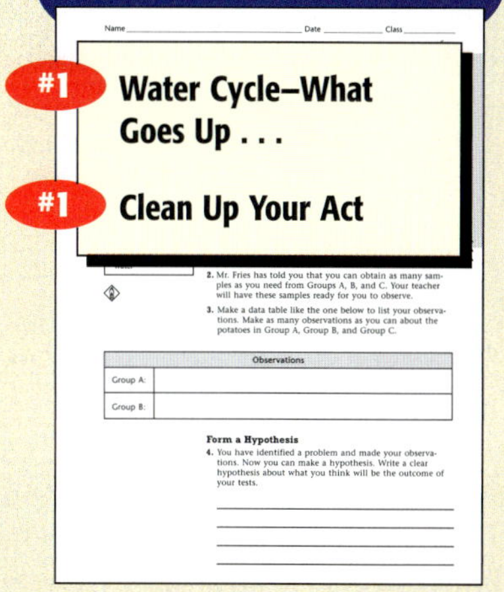
#1 Water Cycle—What Goes Up . . .

#1 Clean Up Your Act

Applications & Extensions

CRITICAL THINKING & PROBLEM SOLVING

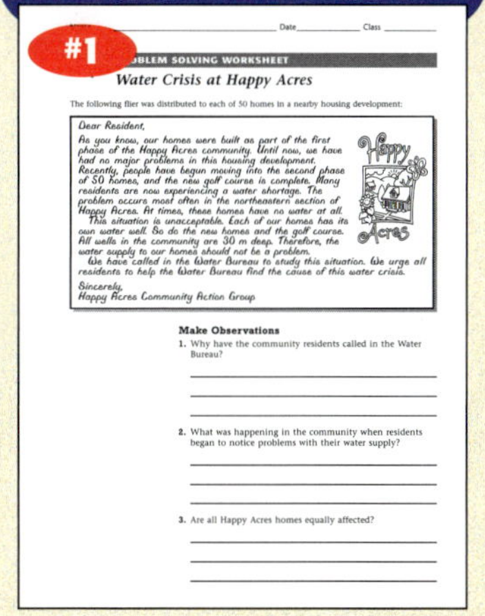
#1
PROBLEM SOLVING WORKSHEET
Water Crisis at Happy Acres

SCIENCE TECHNOLOGY

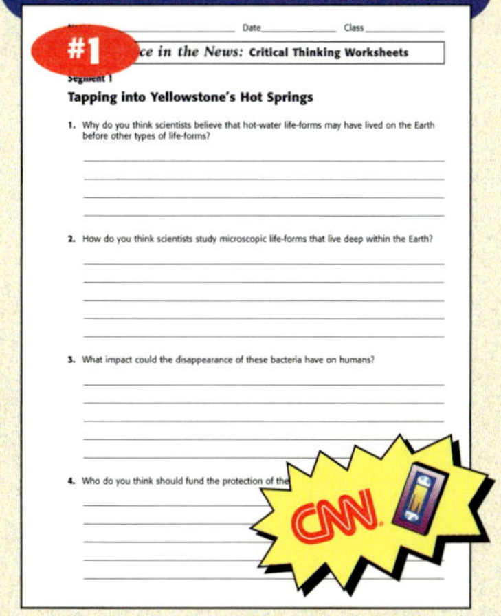
#1
Science in the News: Critical Thinking Worksheets
Tapping into Yellowstone's Hot Springs

EYE ON THE ENVIRONMENT

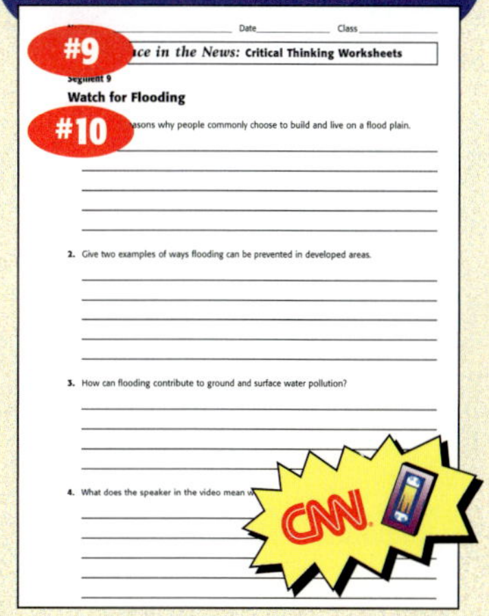
#9
Science in the News: Critical Thinking Worksheets
Watch for Flooding
#10

INTERACTIVE EXPLORATIONS

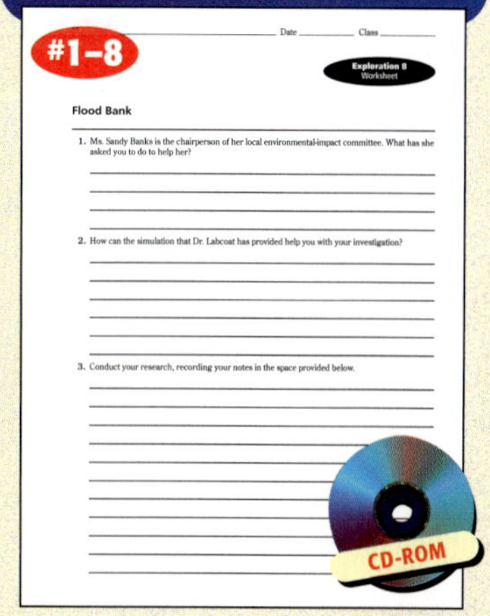
#1–8
Flood Bank
CD-ROM

Chapter Background

Section 1

The Active River

William Morris Davis

William Morris Davis (1850–1934) was a famous American geographer who was the first to propose the erosion cycle. Davis theorized that landscapes are initially uplifted. Streams flow rapidly from the uplifted land, cutting into the landscape. Gradually, the landscape's slope is reduced. Eventually, the landscape changes into an old erosional surface that is fairly flat. Davis's theory is not supported by academics today, however. The process of river erosion is approached today from a systems perspective. Each system is composed of different parts that vary from landscape to landscape.

Stream Flow

Streams have two general types of flow—laminar and turbulent. Laminar flow occurs when the stream load moves in a generally parallel flow. This occurs where the water moves slowly and the channel is smooth. Streams with a greater velocity experience turbulent flow. The stream load generally is rolled, lifted, and bounced along, causing much more erosion in the stream channel.

IS THAT A FACT!

- Geologists have learned that about 300 million years ago, massive rivers may have run down the Appalachian Mountains through river valleys 32 km wide! These rivers drained much of the continent, emptying into the ocean at today's Alabama-Mississippi coastline.

Section 2

Stream and River Deposits

Drainage Patterns

A drainage pattern is the arrangement of river channels in a drainage basin. A drainage pattern is determined by an area's geology and climate. One of the most common patterns, called *dendritic,* is a treelike pattern that forms where rocks and sediments are flat. A *parallel* pattern occurs where there are valleys and ridges. And a *radial* pattern results when streams flow from a central peak, such as a volcanic mountain.

Deltas

The word *delta* was first used by the Greek historian Herodotus to describe the mouth of a river. In the fifth century B.C., Herodotus was traveling in Egypt when he saw the triangular mouth of the Nile River and named the shape after the Greek letter Δ, delta.

Paleoecology at Work

Paleoecologists study the rock and fossil record to understand what ancient environments were like. For example, paleoecologists learned that a great inland sea once covered the Plains states by studying thick layers of sedimentary rock in the area, such as sandstone, shale, and chalk. Such rocks form only when sediments accumulate in calm waters.

IS THAT A FACT!

- Over the last 6,000 years, the Mississippi River delta has shifted from east to west several times. Today the river empties to the east, but scientists think that if left alone the river would change its course and head toward a swampy region called the Atchafalaya. Only massive dams keep the Mississippi on its present course. If the river channel changes, the river may no longer pass through New Orleans.

Section 3

Water Underground

Caves as Shelters

The flow of fresh water underground created caves that were just as important to the development of human civilizations as fertile flood plains. Humans have used caves as shelters for hundreds of thousands of years. Evidence suggests that use of caves for shelter coincides with the first controlled use of fire. Hearths that may be 750,000 years old have been found in the cave of l'Escale, in southeastern France.

- In China, excavations in a cave called Chou-k'ou-tien have yielded 400,000-year-old fossilized remains of *Homo erectus.* Evidence of charred animal bones suggests that the inhabitants may have cooked their food.

IS THAT A FACT!

- The largest cave chamber in the world is the Sarawak chamber, in Malaysia. The chamber is 600 m long and has an average width of 450 m.

Section 4

Using Water Wisely

Aquifers

There are two types of aquifers. The first occurs in *consolidated formations,* those formed from solid rock overlaid with permeable rock that is saturated with water. The second kind of aquifer forms in *unconsolidated formations*—loose sand, soil, and gravel. The amount of water contained in an unconsolidated aquifer depends on how tightly the materials are packed. Because of this, sand and gravel, which are coarse-grained, are usually high-yield aquifers, while formations that are finer-grained tend to hold less water.

Acequias in New Mexico

New Mexico gets only an average of 32.5 cm of rain per year, so water conservation is critical. Because wide irrigation ditches lose a great deal of water to evaporation, New Mexicans have relied for centuries on shallow earthen ditches fed by local rivers to supply growing plants with water. Each ditch, called an *acequia,* provides water to a small area. Acequias allow for water to seep into the ground, minimizing evaporation and allowing water to reach plant roots. Water that isn't absorbed by the soil returns to the river, providing water to people downstream to irrigate their fields.

IS THAT A FACT!

- When ground water is depleted so quickly that the system cannot recharge, there can be dramatic consequences. At Edwards Air Force Base, in California, the aquifer has shrunk so quickly that ground settling has led to sinks and fissures. One of the fissures is about 625 m long!

- A comprehensive 1998 study conducted by the Nature Conservancy warns that continued pollution of the nation's 2,100 rivers and streams threatens approximately 40 percent of fresh-water fish species and two-thirds of mussel species with extinction. In addition, the degradation of our fresh-water sources poses a threat to human health and threatens the country's $16 billion sport-fishing industry.

For background information about teaching strategies and issues, refer to the ***Professional Reference for Teachers.***

CHAPTER 1

The Flow of Fresh Water

Pre-Reading Questions

Students may not know the answers to these questions before reading the chapter, so accept any reasonable response.

Suggested Answers

1. Answers will vary. Accept all reasonable responses. Sample answer: Running water erodes the landscape, redistributing sediment.
2. Erosion is the removal and transport of sediment. Deposition is the dropping off of sediment.

CHAPTER 1

The Flow of Fresh Water

Sections

The Sound Is Deafening

You can hear the thundering roar of Iguaçu (EE gwah SOO) Falls for miles. The Iguaçu River travels more than 500 km across Brazil before it tumbles off the edge of a volcanic plateau in a series of 275 individual waterfalls separated by forested islands. Over the past 20,000 years, erosion has caused the falls to move 28 km upstream. Where will they be 20,000 years from now? In this chapter, you will learn how flowing water shapes Earth's surface.

Pre-Reading Questions

1. What role does water play in shaping the surface of the Earth?
2. What is the difference between erosion and deposition?

internet connect

HRW On-line Resources

go.hrw.com

For worksheets and other teaching aids, visit the HRW Web site and type in the keyword: **HSTDEP**

www.scilinks.com

Use the *sci*LINKS numbers at the end of each chapter for additional resources on the **NSTA** Web site.

www.si.edu/hrw

Visit the Smithsonian Institution Web site for related on-line resources.

www.cnnfyi.com

Visit the CNN Web site for current events coverage and classroom resources.

START-UP Activity

STREAM WEAVERS

How do streams and river systems develop? Do the following activity to find out.

Procedure

1. Begin with a **bucket of sand** and enough **gravel** to fill the bottom of a **rectangular plastic washtub.**
2. Spread the gravel in a layer at the bottom of the washtub. Place 4–6 cm of sand on top of the gravel. Add more sand to one end of the washtub to form a slope.
3. Make a small hole in the bottom of a **paper cup.** Attach the cup to the inside of the tub with a **clothespin.** The cup should be placed at the end that has more sand. Fill the cup with water, and observe the **water** as it moves over the sand. Use a **magnifying lens** to observe features of the stream more closely.
4. Record your observations in your ScienceLog.

Analysis

5. At the start of your experiment, how did the moving water affect the sand?
6. As time passed, how did the moving water affect the sand?
7. Explain how this activity modeled the development of streams. In what ways was it accurate? How was it inaccurate?

3

START-UP Activity

STREAM WEAVERS

MATERIALS

For Each Group:
- bucket of sand
- gravel
- rectangular plastic washtub
- paper cup
- clothespin
- water
- magnifying lens

Teacher's Notes

Students can tilt the tub by placing a block of wood or a book under one end.

Answers to START-UP Activity

5. The moving water cut into the sand, forming a small groove.
6. As time passed, the moving water cut deeper into the sand, creating a wider groove.
7. Accept all reasonable responses. Sample answer: Runoff (water) moves over the land, cutting a gully into the dirt. At first the gully is shallow and narrow, but over time the gully widens and becomes deeper. If there is enough water, the gully eventually becomes a river. Students should note that the model is accurate in that moving water does produce similar landforms. The model is inaccurate for several reasons, including its scale and the fact that the rivers flow over varied terrain, not just uniform sand deposits.

Section 1

Focus

The Active River

This section introduces the water cycle and discusses the role that rivers play in the movement of fresh water. Students will learn that rivers are changing, dynamic systems that continually shape the land. The section discusses the factors that contribute to rates of stream erosion and concludes with a discussion of the life cycle of rivers.

Bellringer

Have students write a paragraph about an imaginary canoe trip down a river. They could choose the river described in the chapter opener or another river of their choice. Ask students to describe the river's features and hypothesize how those features formed.
Sheltered English

1) Motivate

Activity

Discuss the photograph of the Grand Canyon shown in **Figure 1.** Help students compare the size of the canyon with the size of the Empire State Building, which is 381 m tall. Ask students to calculate how many Empire State Buildings it would take to create a stack as deep as the Grand Canyon. (1.6 km deep = 1,600 m; 1,600 ÷ 381 = 4.2)

The Grand Canyon is 29 km wide at its widest point. How many Empire State Buildings laid end-to-end would fit in the canyon's widest point? (29 km = 29,000 m; 29,000 ÷ 381 = 76)

Section 1 — Reading Warm-Up

Terms to Learn

erosion
water cycle
tributary
drainage basin
divide
channel
load

What You'll Do

- Illustrate the water cycle.
- Describe a drainage basin.
- Explain the major factors that affect the rate of stream erosion.
- Identify the stages of river development.

The Active River

You are probably familiar with the Grand Canyon, shown in **Figure 1.** But did you know that about 6 million years ago, the area now known as the Grand Canyon was nearly as flat as a pancake? The Colorado River cut down into the rock and formed the Grand Canyon over millions of years by washing billions of tons of soil and rock from its riverbed. This process is a type of *erosion*. **Erosion** is the removal and transport of surface material, such as rock and soil. Rivers are not the only agents of erosion. Wind, rain, ice, and snow can cause erosion as well.

Because of erosion caused by water, the Grand Canyon is now about 1.6 km deep and 446 km long. In this section, you will learn about stream development, river systems, and the different factors that affect the rate of stream erosion.

Figure 1 *The Grand Canyon is located in northwestern Arizona. It formed over millions of years as running water eroded rock and soil. In some places the canyon is 29 km wide.*

Water, Water Everywhere

Have you ever wondered how rivers keep flowing and where rivers get their water? The water cycle answers these and other questions. The **water cycle,** shown on the next page, is the continuous movement of water from water sources, such as lakes or oceans, into the air, onto land, into the ground, and back to the water sources.

4

Math Skills Worksheet "Checking Division with Multiplication"

internetconnect

SCLINKS NSTA
TOPIC: The Grand Canyon
GO TO: www.scilinks.org
***sci*LINKS NUMBER:** HSTE255

The Water Cycle

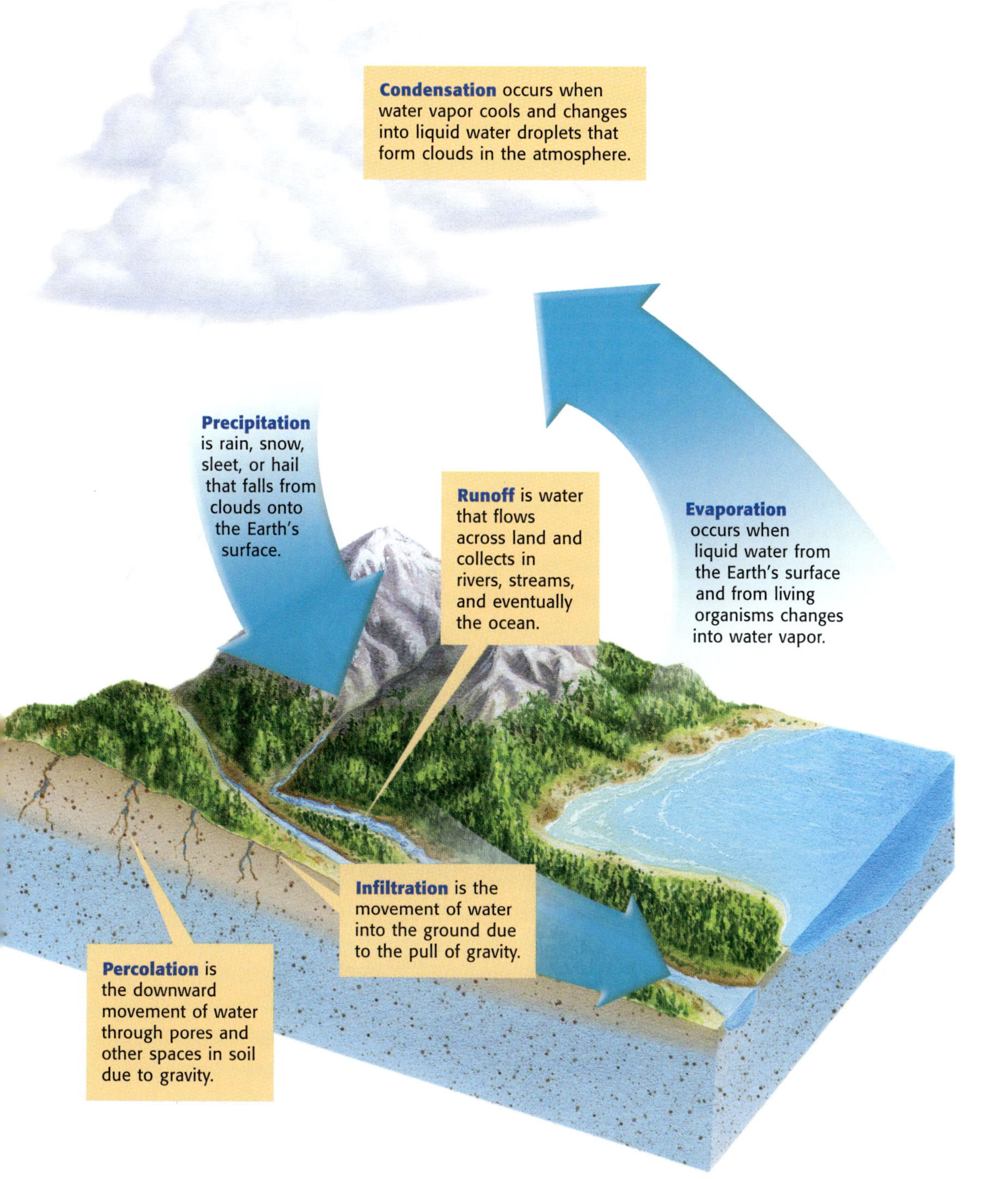

2 Teach

DEMONSTRATION

Add a teaspoon of brightly colored tempera paint to about half a cup of soil. Place the soil in a funnel lined with filter paper, and place the funnel over a large jar. Tell students that the tempera paint represents nutrients in the soil. Ask students to predict what will happen to the nutrients in the soil when it rains. Demonstrate by slowly pouring water into the funnel. Discuss the important role that rivers play in distributing soil nutrients.

MEETING INDIVIDUAL NEEDS

Learners Having Difficulty

Have students make a field guide for rivers. Suggest that they include photographs of streams and rivers and write captions to describe each photo. The captions should incorporate terms such as *gradient, erosion, load, channel,* and *meanders*. Finally, have them hypothesize about the river's speed, load, and erosional capacity. Sheltered English

Teaching Transparency 141 "The Water Cycle"

Directed Reading Worksheet Section 1

IS THAT A FACT!

Under average conditions, the Mississippi River carries about 17,000 m^3 of water every second. A small carry-on suitcase is about 0.03 m^3, so watching the river go by on an average day is equivalent to watching about 566,666 suitcases pass by you every second!

MISCONCEPTION ALERT

Be sure students realize that water evaporates from the soil and vegetation as well as from rivers and other bodies of water.

Answers to Activity

Tributaries: Musselshell, Yellowstone, Badlands, Heart, Grand, Moreau, Cheyenne, White, Niobrara, Elkhorn, Loup, Platte, Kansas, Ohio, Arkansas, Red

Cities: Bismarck, Omaha, Kansas City, Saint Louis, Memphis, Baton Rouge, New Orleans

Distance: approximately 3,900 km

Hint: Measure the distance with a string, then find the actual distance traveled by using the map's scale.

Group Activity

Have pairs of students use a map or atlas to create a poster that illustrates the river systems and drainage basins of North America. Encourage them to use arrows to indicate the direction of flow in major rivers and draw the divides that separate each drainage basin. Working independently, students could make a map that shows the river systems of your county or state.

Real-World Connection

Meteorologists recognize the power of moving water, and they issue special warnings when there is a danger of floods. Ask students to guess how much moving water it takes to knock an adult off his or her feet:

15 cm, 30 cm, or 1 m. (15 cm)

Ask them how deep the water would have to be to sweep away a car:

more than 2 m, at least 1.5 m, or less than 1 m. (less than 1 m)

Demonstrate these water depths using a meterstick, and caution students to be careful around flowing water.

Activity

Imagine that you are planning a rafting trip down the Missouri River to the Mississippi River. On a map of the United States, trace the route of your trip from the Rocky Mountains in Montana to the mouth of the Mississippi River, in Louisiana. What major tributaries would you travel past? What cities would you pass through? Mark them on the map. How many kilometers would you travel on this trip?

TRY at HOME

River Systems

Look at the pattern of lines on the palm of your hand. Notice how some of the smaller lines join together to form larger lines. Now imagine those lines are rivers and streams. The smaller lines would be the streams and tributaries and the larger lines would be rivers. **Tributaries** are smaller streams or rivers that flow into larger ones. Like the network of lines on the palm of your hand, streams and rivers make up a network on land. This network of streams and rivers is called a river system and it drains an area of its runoff.

Drainage Basins River systems are divided into regions known as drainage basins. A **drainage basin,** or *watershed,* is the land drained by a river system, which includes the main river and all of its tributaries. The largest drainage basin in the United States is the Mississippi River basin. It has hundreds of tributaries that extend from the Rocky Mountains, in the West, to the Appalachian Mountains, in the East.

The map in **Figure 2** shows that the Mississippi River drainage basin covers more than one-third of the United States. Other major drainage basins in the United States are the Columbia, Rio Grande, and Colorado River basins.

Divides Drainage basins are separated from each other by an area called a **divide.** A divide is generally an area of higher ground than the basins it separates. On the map below, you can see that the Continental Divide is a major divide in the United States. On which side do you live?

BRAIN FOOD

The Amazon River basin, in South America, is the world's largest drainage basin. It has an area of about 6 million square kilometers. That's almost twice as big as the United States' largest drainage basin, the Mississippi River basin!

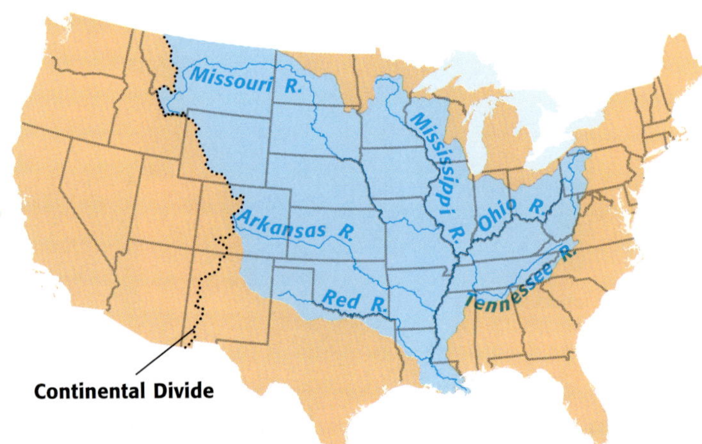

Figure 2 *The Continental Divide runs through the Rocky Mountains. It separates the drainage basins that flow into the Atlantic Ocean and the Gulf of Mexico from those that flow into the Pacific Ocean.*

Science Bloopers

In 1935, a dam was constructed in California to prevent sediment from filling Gibraltar Reservoir. This dam created Mono Reservoir. The watershed above Mono Reservoir was burned by forest fires, and people worried that sediment would fill Gibraltar Reservoir before plants had a chance to regrow and hold the soil. Unfortunately, the next 2 years saw record rainfall. Sediment completely filled Mono Reservoir and half-filled Gibraltar Reservoir.

Stream Erosion

As a stream forms, it erodes soil and rock to create a channel. A **channel** is the path that a stream follows. At first, stream channels are small and steep. As more rock and soil are transported downstream, the channels become wider and deeper. When streams become longer, they are referred to as rivers. Have you ever wondered why some streams flow faster than others?

Gradient The stream shown in **Figure 3** is flowing down a steep mountain side. This stream has a high gradient. *Gradient* is the measure of the change in elevation over a certain distance. A high gradient gives a stream or river more energy to erode rock and soil. A river or stream with a low gradient, such as the one in **Figure 4,** has less energy for erosion.

Discharge The amount of water a stream or river carries in a given amount of time is called *discharge*. The discharge of a stream increases when a major storm occurs or when warm weather rapidly melts snow. As the stream's discharge increases, its erosive energy, speed, and load increase.

MATHBREAK

Calculating a Stream's Gradient

If a river starts at an elevation of 4,900 m and travels 450 km downstream to a lake that is at an elevation of 400 m, what is the stream's gradient?

Figure 3 *A mountain stream flows rapidly and has more erosive energy.*

Figure 4 *A river on a flat plain flows slowly and has less erosive energy.*

7

IS THAT A FACT!

Most river water comes from rainfall and melted snow that flows down from mountains. Why do some rivers continue to flow during a severe drought? These rivers are probably lower than the water table, so their flow is maintained by seepage from ground water.

WEIRD SCIENCE

Stream piracy occurs when one river "steals" another. If the land dividing two streams is eroded, one stream can "capture" the headwaters of the other. The eroding stream eventually pirates all of the other stream's water.

Answer to MATHBREAK

4,900 m − 400 m = 4,500 m

$\frac{4,500 \text{ m}}{450 \text{ km}}$ = 10 m/km

Discussion

Have students think about watering a garden using a hose. Ask whether they would spray the water at full blast or gently. Would they lay the hose on the ground or hold it at waist level? Refer to this example as you work through the concepts of erosion, gradient, and discharge. Sheltered English

CONNECT TO PHYSICAL SCIENCE

Water flowing downstream can be compared to a rock tumbling down a mountainside. Water's potential energy is constantly being converted into kinetic energy. The reservoir of a hydroelectric dam has tremendous potential energy; when the gates are open, that potential energy is converted into kinetic energy that spins a turbine. The turbine then generates electrical energy. Ask students to come up with other devices that use the potential energy of water. (Examples include a flush toilet and a water tank.)

MISCONCEPTION ALERT

When water flows down a river channel, it does not flow straight down the channel. Water flowing in a river or stream channel moves in a helical or corkscrew motion. This motion causes erosion on the bank where the water is rising and causes deposition on the bank where the water is falling. The helical flow of water helps explain the formation of bends or meanders in the channel.

2 Teach, continued

Group Activity

Assign each group of students one of the world's major rivers. Have each group prepare a poster on the river indicating its headwaters and mouth. The poster should also include major tributaries. Have students indicate a change in the river's gradient by shading the area with colored pencils. Encourage students to discover interesting facts about their river, such as the ecosystem the river supports and the river's effects on human populations living near it.

Cooperative Learning

Stream Load To help students learn the differences between the types of loads that streams carry, engage groups of four in this activity. Have groups fill a plastic jar three-quarters full of water. Provide a few small pebbles, a quarter cup of soil, and 3 tbsp of salt for each group. Have students choose which material best represents a stream's bed load (the pebbles), its suspended load (the soil), and its dissolved load (the salt).

Have students add all three materials to the jar and then shake it carefully to simulate a stream's load. After the contents have been thoroughly mixed, ask students to hypothesize how they could remove each material from the jar. (The pebbles settle to the bottom and can be picked out easily. If the water remains still long enough, the sediment will settle to the bottom of the jar. The salt can be removed through evaporation.)

Answer to Self-Check

If a river slowed down, the suspended load would be deposited.

Load The materials carried in a stream's water are collectively called the stream's **load.** The size of the particles in the stream's load is affected by the stream's speed. Fast-moving streams can carry large particles. The load also affects the stream's rate of erosion. Rocks and pebbles bounce and scrape along the bottom and sides of the stream bed. The illustration below shows the three ways a stream can carry its load.

Self-Check

What would happen to a suspended load if the river slowed down? *(See page 136 to check your answer.)*

Homework

Writing Have students prepare an *Encyclopedia of Rivers* with a full definition of each new term in the section and at least one illustration for each term.

Science Humor

Q: What is a flood?

A: a river that's too big for its bridges

The Stages of a River

In the early 1900s, William Morris Davis developed a model that identified the stages of river development. According to this model, rivers evolve from a youthful stage to an old-age stage. Davis believed that all rivers erode in the same way and at the same rate. Today, however, scientists support a different model that considers the effects of a river's environment on stream development. For example, because different material erodes at different rates, one river may develop more quickly than another river. Many factors, including climate, gradient, and load, influence the development of a river. Although scientists no longer use Davis's model to explain river development, they still use many of his terms to describe a river. Remember, these terms do not tell the actual age of a river. Instead, they are used to describe the general characteristics of the river.

Figure 5 *This youthful river is located in Yellowstone National Park in Wyoming. The rapids and falls are located where the river flows over hard, resistant rock.*

Youthful Rivers A youthful river, like the one shown in **Figure 5,** erodes its channel deeper rather than wider. The river flows quickly because of its steep gradient. Its sides and channel are steep and straight. The river tumbles over rocks in rapids and waterfalls. Youthful rivers have few tributaries.

Mature Rivers A mature river, as shown in **Figure 6,** erodes its channel wider rather than deeper. The gradient of a mature river is not as steep as that of a youthful river, and there are fewer falls and rapids. A mature river is fed by many tributaries, and because of its good drainage, it has more discharge than a younger river.

Figure 6 *A mature river, such as this one in Peru, begins to curve back and forth. The bends in the river's channel are called* meanders.

3 Extend

Reading Strategy

Prediction Guide Write "Young," "Mature," "Old," and "Rejuvenated" on the board. Have students brainstorm about concepts, words, or images related to these terms, and write their suggestions below the terms. Next add the dictionary definitions of the terms. Then discuss how the terms might describe a river. Have students write their ideas in their ScienceLog and then compare their ideas with the descriptions in the text.

Sheltered English

Independent Practice

PORTFOLIO

Illustrating River Stages As students read the section on the stages of a river, have them illustrate the stages in their ScienceLog. Encourage students to draw a cross section of the river channel and valley. Students should label the parts of their diagrams and write a brief description of each stage of river development.

Cross-Disciplinary Focus

Writing

History In the eighteenth and nineteenth centuries, the Mississippi River served as an important trade route for the Midwest. Riverboats carried people and goods up and down the Mississippi River. The riverboats carried resources such as grains, meat, and animal skins down the river. Have interested students research how the Mississippi River influenced the history of the Midwest or write a short essay on the history of a city built on a Midwestern river, such as Saint Louis, New Orleans, or Kansas City. They should concentrate on how the river was important to the city's growth and development. Students should also find out how the Mississippi River is used today.

4 Close

Quiz

1. How does rainfall or snowmelt affect a river's discharge and its ability to cause erosion? (Rainfall and snowmelt increase a river's discharge and the amount of erosion it can cause.)
2. Explain the three types of materials carried by a river. (bed load: large materials that roll or bounce along a riverbed; suspended load: materials that float suspended in a river; dissolved load: materials dissolved in a river)

Alternative Assessment

Display the Teaching Transparency listed below, and discuss the features labeled. Then divide students into groups, and have them use modeling clay to make a model of a river, incorporating the concepts they have learned in this chapter. Students can use pins with attached labels to indicate river features.

Teaching Transparency 75
"River Features"
LINK TO LIFE SCIENCE

internet**connect**
SCILINKS NSTA
TOPIC: Rivers and Streams
GO TO: www.scilinks.org
***sci*LINKS NUMBER:** HSTE260

Figure 7 *This old river is located in New Zealand.*

Old Rivers An old river has a low gradient and extremely low erosive power. Instead of widening and deepening its banks, the river deposits sediment in its channel and along its banks. Old rivers, like the one in **Figure 7,** are characterized by wide, flat *flood plains*, or valleys, and more meanders. Also, an older river has fewer tributaries than a mature river because the smaller tributaries have merged.

Figure 8 *This rejuvenated river is located in Canyonlands National Park, Utah.*

Rejuvenated Rivers Rejuvenated rivers occur where the land is raised by the Earth's tectonic forces. When land rises, the river's gradient becomes steeper. The increased gradient of a rejuvenated river allows the river to cut more deeply into the valley floor, as shown in **Figure 8.** Steplike *terraces* often form on both sides of a stream valley as a result of rejuvenation. Terraces are nearly flat portions of the landscape that end at a steep cliff.

internet**connect**
SCILINKS NSTA
TOPIC: Rivers and Streams
GO TO: www.scilinks.org
***sci*LINKS NUMBER:** HSTE260

SECTION REVIEW

1. How does the water cycle help to develop river systems?
2. Describe a drainage basin.
3. What are three factors that affect the rate of stream erosion?
4. **Summarizing Data** How do youthful, mature, and old rivers differ?

Answers to Section Review

1. Answers will vary. Accept all reasonable responses.
2. A drainage basin is the land drained by a river system, including the main river and all of its tributaries.
3. Three factors that affect the rate of stream erosion are a stream's gradient, its discharge, and its load.
4. Youthful rivers are steep and straight, with fewer tributaries. Mature rivers have more curves, tributaries, and discharge than youthful rivers, as well as a lower gradient. Old rivers have less erosive power than than mature rivers do, but these tributaries are larger and they meander more.

SECTION 2 READING WARM-UP

Terms to Learn

deposition
alluvium
delta
alluvial fan
flood plain

What You'll Do

- Describe the different types of stream deposits.
- Explain the relationship between rich agricultural regions and river flood plains.

Stream and River Deposits

You have learned that flowing rivers can pick up and move soil and rock. Sooner or later, this material must be deposited somewhere. **Deposition** is the process by which material is dropped, or settles. Imagine a mud puddle after a rainy day. If the water is not disturbed, the soil particles will eventually settle and the muddy water will become clear again. Deposition also forms and renews some of the world's most productive soils. People who live in the lower Mississippi River valley, for example, depend on the river to bring them new, fertile soil.

Deposition in Water

After rivers erode rock and soil, they deposit the rock and soil downstream. Rock and soil deposited by streams is called **alluvium.** Alluvium is dropped at places in a river where the speed of the current decreases. Take a look at **Figure 9** to see how this type of deposition occurs.

Figure 9 *This model illustrates erosion and deposition at a bend, or meander, of a river.*

WEIRD SCIENCE

The Okavango River, in Africa, flows approximately 1,600 km through Angola, Namibia, and Botswana before it empties into the middle of the Kalahari Desert, where it evaporates! The river's alluvial fan provides a haven for the desert's plants and animals.

internet connect

SCI LINKS NSTA
TOPIC: Stream Deposits
GO TO: www.scilinks.org
***sci*LINKS NUMBER:** HSTE263

SECTION 2

Focus

Stream and River Deposits

This section describes the variety of ways that a river's load can be deposited. Students will learn why different patterns of deposition occur, and they will explore the connections between river deposits and agriculture.

Bellringer

Post the following question on the board or overhead projector:

Even though flooding along rivers is potentially harmful, many farmers have traditionally located their farms near rivers. Why? (The flood waters carry rich sediments that contribute to the land's fertility.)

1) Motivate

DEMONSTRATION

Modeling Deposition To show students how moving water separates sediment into different-sized particles, place a handful of soil, sand, and small gravel in a large plastic jar filled with water. Shake the jar vigorously, and then place it in on your desk. Ask students to describe what happens to the different-sized particles in the jar as time passes. Ask students to compare how the different-sized particles settled. Sheltered English

Directed Reading Worksheet Section 2

2 Teach

BRAIN FOOD

Ask students to examine a map of the United States and to note that rivers often serve as state borders. For instance, the Ohio River is the border between Indiana and Kentucky. But meanders can sometimes shift the course of the river. Does the state border then change? To prevent feuding between states, the courts have declared that state boundaries do not change when a river shifts.

COOPERATIVE LEARNING

PORTFOLIO

Gold Rush Between 1870 and 1898, many prospectors discovered gold in Canada's Yukon Territory. Dreams of getting rich overnight lured people to the Klondike River area in vast numbers.

Have students research the Yukon gold rush. Divide the class into three groups. Have Group 1 prepare a presentation on how the gold deposits formed. Have Group 2 prepare a presentation on how the gold ore became placer deposits in the Klondike River. Have Group 3 prepare a presentation on the geography of the Yukon's river systems.

Students can also find out about life in the gold fields and present their findings using models or posters.

Reinforcement Worksheet
"Fresh Water in the United States"

Figure 10 *Miners rushed to California in the 1850s to find gold. They often found it in the bends of rivers in placer deposits.*

Heavy minerals are sometimes deposited at places in a river where the current slows down. This kind of alluvium is called a *placer deposit*. Some placer deposits contain gold, as **Figure 10** shows. During the California gold rush, which began in 1849, many miners panned for gold in the placer deposits of rivers.

Designing a Delta The current also slows when a river empties into a large body of water, such as a lake or an ocean. Much of the river's load may be deposited where the river reaches the large body of water, forming a fan-shaped deposit called a **delta.** In **Figure 11** you can see an astronaut's view of the Nile Delta. A delta usually forms on a flat surface and consists mostly of mud. These mud deposits form new land, causing the coastline to grow.

Figure 11 *Alluvium is dropped at the mouth of the Nile River, forming a delta.*

If you look back at the map of the Mississippi River drainage basin in Figure 2, you can see where the Mississippi River flows into the Gulf of Mexico. This is where the Mississippi Delta has formed. Each of the fine mud particles in the delta began its journey far upstream. Parts of Louisiana are made up of particles that were transported from as far away as Montana, Minnesota, Ohio, and Illinois.

Astronomy CONNECTION

The remains of an ancient riverbed have been discovered on Mars. Satellite images show the deposits of stream channels, which indicate that liquid water once existed on the now dry and frozen planet.

Self-Check

What is one factor that causes the current of a river to slow? *(See page 136 to check your answer.)*

12

Answer to Self-Check

Answers will vary. A river might slow where there is a bend, when the gradient decreases, or where the river empties into a large body of water.

Homework

Investigate Your Area Have students find out how your city or county is working to control flooding and erosion along area streams and rivers. Have them write a brief report on a watershed project in their area. You may consider inviting a watershed engineer to speak to the class.

Deposition on Land

When a fast-moving mountain stream flows onto a flat plain, the stream slows down. As the stream slows down, it deposits alluvium where the mountain meets the flat plain, forming an alluvial fan, such as the one shown in **Figure 12.** **Alluvial fans** are fan-shaped deposits that form on dry land.

During periods of high rainfall or rapid snowmelt, a sudden increase in the volume of water flowing into a stream can cause the stream to overflow its banks, flooding the surrounding land. This land is called a **flood plain.** When a stream floods, a layer of alluvium is deposited across the flood plain. Each flood adds another layer of alluvium.

Fatal Flooding Flood plains are very rich farming areas because periodic flooding brings new soil to the land. However, flooding can cause extensive property damage. Much farming activity takes place in the Mississippi River valley, a large flood plain with very rich soil. When the Mississippi River flooded in 1993, however, farms were abandoned and whole towns had to be evacuated. The flood was so huge that it caused damage in nine Midwestern states. **Figure 13** shows an area that was flooded just north of St. Louis, Missouri.

Figure 12 *An alluvial fan, such as this one from the Sierra Nevada, in California, forms when an eroding stream changes rapidly into a depositing stream.*

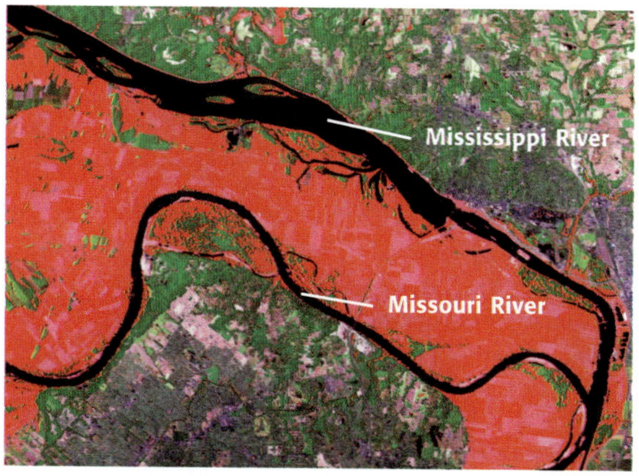

Figure 13 *The normal flow of the Mississippi River and Missouri River is shown in black. The area that was flooded when both rivers spilled over their banks in 1993 is shaded red.*

SECTION REVIEW

1. What happens to a river's flow that causes alluvium to be deposited?
2. How are alluvial fans and deltas similar? How are they different?
3. Explain why flood plains are good farming areas.
4. **Identifying Relationships** What factors increase the likelihood that alluvium will be deposited?

internet**connect**

SCILINKS NSTA

TOPIC: Stream Deposits
GO TO: www.scilinks.org
***sci*LINKS NUMBER:** HSTE263

13

DEBATE

Flood Management The flooding of the Mississippi River in 1993 spurred a great debate. Some people blamed the severity of the flood on development projects that had destroyed wetlands and the natural flood plain. They argued that parts of the flood plain should be restored to their natural state. Others argued that if the natural habitat were restored, people living and working in the flood plain would be displaced. Others wondered if insurance companies should pay to rebuild houses that were located in flood plains.

Have two groups research the controversy and defend their positions in a class debate.

3 Close

Quiz

1. A river runs down a rapids, eases through a valley for about 3 km, then tumbles down a waterfall into a lake. Where along this path would you most likely find a placer deposit? Why? (in the valley, because heavy minerals are sometimes deposited where currents are slow)
2. Define *flood plain*. (land that is flooded when a river overflows its banks)

ALTERNATIVE ASSESSMENT

Have students use modeling clay to create a model of a delta and an alluvial fan. Ask them to also model the immediate environment around each river feature and use pins with labels to identify the processes involved in the formation of each feature.

Answers to Section Review

1. The slowing of a river's flow causes alluvium to be deposited.
2. Deltas and alluvial fans are both fan-shaped deposits formed when a stream slows. Deltas form as rivers enter larger bodies of water, and alluvial fans form on dry land.
3. Flood plains form rich farmland because flood waters periodically deposit new soil on the land.
4. Factors that reduce a river's speed increase the likelihood that alluvium will be deposited.

SECTION 3

Focus

Water Underground

In this section, students will learn what ground water is and where it is found. The section discusses the formation of aquifers and their importance to agriculture and human populations. Students will learn how wells and springs bring water from aquifers to the surface. The section concludes with a discussion of how the movement of ground water forms caves and sinkholes.

Bellringer

Pose the following scenario to students on the board or overhead projector:

A family lives 50 km from the nearest stream or lake and gets water from a well. Where does the water in the well come from? (It comes from water stored underground.)

1 Motivate

DEMONSTRATION

Layer an aquarium with gravel, sand, and potting soil until it is three-quarters full. Pack the material firmly. Add water until you can clearly see areas of aeration and saturation. Ask students to compare this model to **Figure 14.** Using a marker, draw a line for the water table on the glass. Introduce the terms *zone of saturation* and *zone of aeration*, and discuss the difference between the two. Sheltered English

Teaching Transparency 142
"The Water Table"

SECTION 3 READING WARM-UP

Terms to Learn

ground water
water table
aquifer
porosity
permeability
recharge zone
artesian spring

What You'll Do

- Identify and describe the location of a water table.
- Describe the characteristics of an aquifer.
- Explain how caves and sinkholes form as a result of erosion and deposition.

Water Underground

Although we can see surface water in streams and lakes, there is a lot of water flowing underground that we cannot see. The water located within the rocks below the Earth's surface is called **ground water.** Ground water not only is an important resource but also plays an important role in erosion and deposition.

Location of Ground Water

Surface water seeps underground into the soil and rock. Earth scientists divide this underground area into two zones. The upper zone, called the *zone of aeration,* usually is not completely filled with water. The rock and soil that make up this zone are filled with water only immediately after a rain. Farther down, the water accumulates in an area called the *zone of saturation*. Here the spaces between the rock particles are filled with water.

These two zones meet at an underground boundary known as the **water table,** as shown in **Figure 14.** The water table rises during wet seasons and drops during dry seasons. In wet regions the water table can be just beneath the soil's surface or at the surface. But in deserts the water table may be hundreds of meters underground.

Figure 14 *The water table is the upper surface of the zone of saturation.*

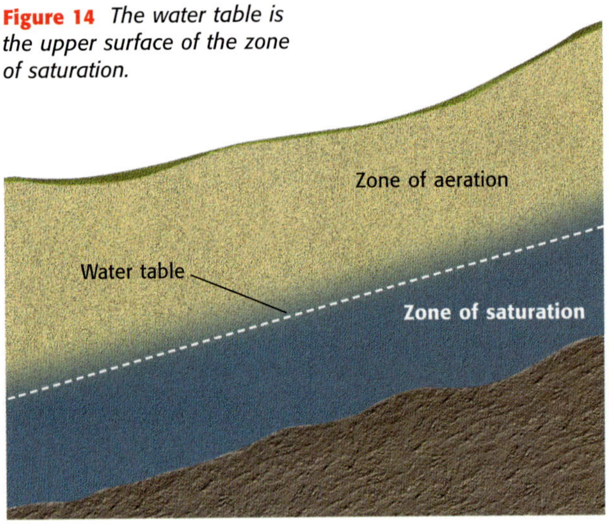

Aquifers

Some types of rock can hold large quantities of water, while other types can hold little or no water. A rock layer that stores and allows the flow of ground water is called an **aquifer.**

To qualify as an aquifer, a rock layer must be *porous*, or contain open spaces. A rock's **porosity** is the amount of open space between individual rock particles. The rock layer must also allow water to pass freely through it, from one pore to another. If the pores are connected, ground water can flow through the rock layer. A rock's ability to let water pass through it is called **permeability.** A rock that tends to stop the flow of water is impermeable.

14

Multicultural CONNECTION

The Maya, one of the early civilizations of the Americas, believed that rain clouds formed in caves and then rose to the sky. Elaborate ceremonies were performed in caves for the rain god. Have interested students find out more about how caves and rituals were important to early Mesoamerican civilizations.

IS THAT A FACT!

Geologists estimate that aquifers hold 50 million cubic kilometers of water worldwide. There is about 20 times more water underground than in all the rivers, lakes, and the atmosphere.

Aquifer Geology and Geography The best aquifers are usually formed of sandstone, limestone, or layers of sand and gravel. Some aquifers cover large underground areas and are an important source of water for cities and agriculture. The map in **Figure 15** shows the location of aquifers in the United States.

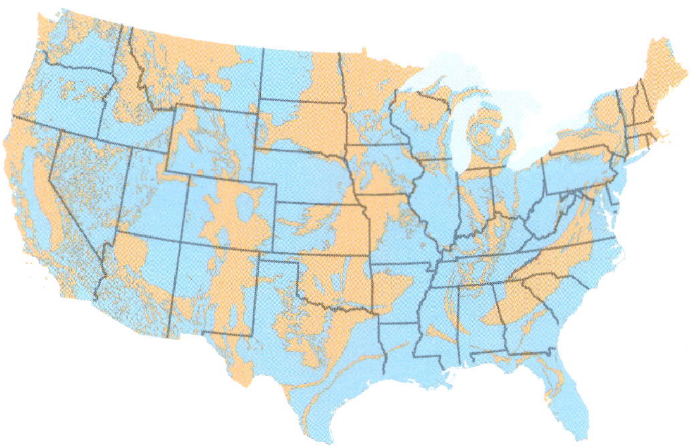

Figure 15 Aquifers in the Continental United States

Recharge Zones Like rivers, aquifers are dependent on the water cycle to maintain a constant flow of water. The ground surface where water enters an aquifer is called the **recharge zone.** The size of the recharge zone varies depending on how permeable rock is at the surface. In an area that contains a permeable rock layer, the water can seep down into the aquifer. In areas where the aquifer is confined on top by an impermeable rock layer, the recharge zone is restricted to areas where there is a permeable rock layer.

Springs and Wells

Ground-water movement is determined by the slope of the water table. Just like surface water, ground water tends to move downslope, toward lower elevations. If the water table reaches the Earth's surface, water will flow out from the ground, forming a *spring*. Springs are an important source of drinking water. Lakes form in low areas, where the water table is higher than the Earth's surface.

QuickLab

Degree of Permeability

1. Obtain five **plastic-foam cups.**
2. Fill one cup halfway with **soil,** such as garden soil. Pack the soil.
3. Fill a second cup halfway with **sand.** Pack the sand.
4. Poke 5 to 7 holes in the bottom of each cup with a sharpened **pencil.**
5. Fill a third cup with **water.** Hold one of the remaining empty cups under the cup filled with soil. Pour the water into the top cup.
6. Allow the cup to drain for 45 seconds, and then put the cup aside (even if it is still draining). Put the cup filled with water aside.
7. Repeat steps 5 and 6 with the cup of sand. Compare the volumes of the two cups of water. The cup that allowed the most water to pass holds the more permeable sediment.

TRY at HOME

A mud pie the size of a house—where would you see something like that? Turn to page 30 to find out.

WEIRD SCIENCE

Only recently have scientists discovered how much fresh water flows into the oceans from ground water. When the tide comes in, sea water seeps into the sediments of coastal land, mixing with the ground water. When the tide recedes, the water mixture is drawn out to sea; the ground water is recharged and the cycle repeats. In addition, many fresh-water aquifers drain directly into the ocean. These findings are significant because pollution in ground water can seriously threaten the oceans.

2 Teach

QuickLab

MATERIALS

- 5 plastic-foam cups
- soil
- sand
- pencil
- water

Teacher Notes: There are no safety precautions required for this experiment. Cover the work area with plastic, and have students wear lab aprons and plastic gloves.

REAL-WORLD CONNECTION

Investigate Your Area Ground water is a drinking water source for about 60 percent of the population of the United States. Encourage students to find out where their drinking water comes from. Students can do research in local newspapers to learn about issues affecting the quality of their drinking water and prepare a poster illustrating the path that local ground water follows from a rain cloud to a faucet. If there is time, arrange for a tour of a local water treatment plant.

MISCONCEPTION ALERT

Students may think that ground water creates vast underground lakes or rivers. This happens only rarely. Most ground water moves through spaces in rock and soil similar to the way water is stored in a sponge.

Directed Reading Worksheet Section 3

2 Teach, continued

Using the Figure

Guide students through **Figures 16** and **17.** List the labels on the board. Help students define each term and explain that artesian springs are driven by hydraulic pressure. Artesian springs occur in areas where the water table is above the outlet for the spring. If students have ever siphoned a liquid, they have applied the principle that causes artesian springs to flow. Challenge students to demonstrate an artesian spring using a water-filled container and a length of tubing. Sheltered English

BRAIN FOOD

Much of Australia's ground water is stored in artesian formations, such as those fed by a vast underground rock formation called the Great Artesian Basin. Unfortunately, most of the water is too salty for people to drink or use for irrigation. Have students find out how this water could be processed in a desalination plant.

Answer to Self-Check

The impermeable rock layer in the aquifer traps the water in the permeable layer below. This creates the pressure needed to form an artesian spring.

Teaching Transparency 143
"Artesian Formation"
"The Water Table and Wells"

Artesian Springs A sloping layer of permeable rock sandwiched between two layers of impermeable rock is called an *artesian formation*. The permeable rock is an aquifer, and the top layer of the impermeable rock is called a *cap rock*, as shown in **Figure 16.** Artesian formations are the source of water for **artesian springs.** Artesian springs are springs that form where cracks occur naturally in the cap rock and the pressurized water in the aquifer flows through the cracks to the surface. Artesian springs are sometimes found in deserts, where they are often the only source of water.

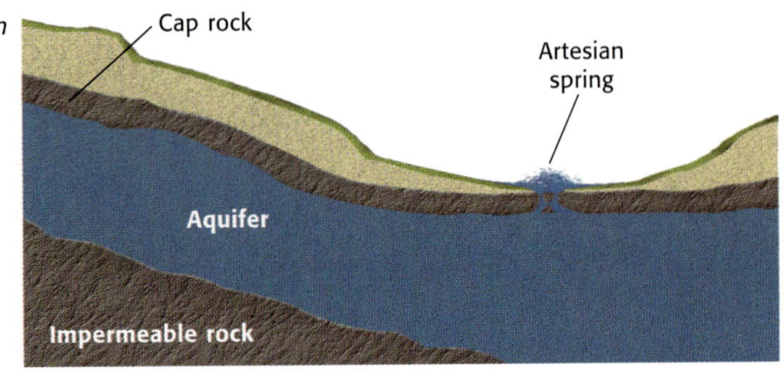

Figure 16 *Artesian springs form when water from an aquifer flows through cracks in the cap rock of an artesian formation.*

Wells A *well* is a human-made hole that is deeper than the level of the water table; therefore, wells fill with ground water, as shown in **Figure 17.** If a well is not deep enough, it will dry up when the water table falls below the bottom of the well. Also, if too many wells in an area remove ground water too rapidly, the water table will drop and all the wells will run dry.

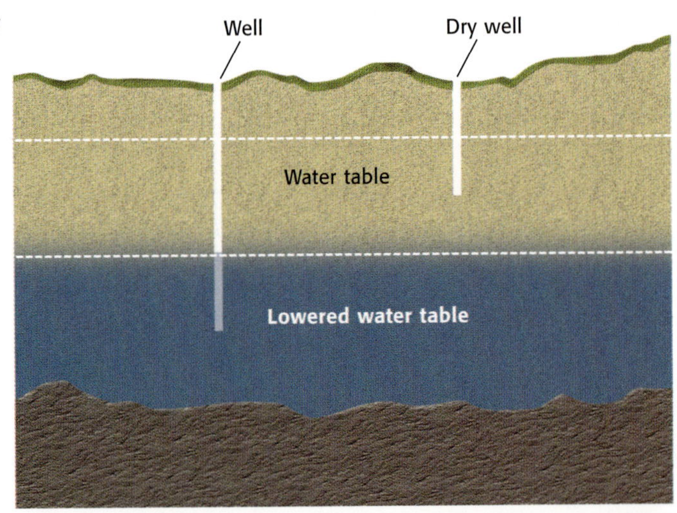

Figure 17 *A good well is drilled deep enough so that when the water table drops, the well still contains water.*

Self-Check

Why is it important that there is a layer of impermeable rock in an artesian formation? *(See page 136 to check your answer.)*

16

CONNECT TO LIFE SCIENCE

Elephants that live in the desert of Namibia may travel 4 days to find a water hole. Once there, they might have to dig almost a meter into the sand to reach clear water. Antelopes, such as springbok and eland, and other desert-dwelling animals depend on the elephants to penetrate the water table. Have students research ways in which desert plants and animals have adapted to regions with little surface water.

Underground Erosion and Deposition

Unlike a river, which erodes its banks when water moves over rock and soil, ground water erodes certain types of rock by dissolving the rock. Most of the world's caves formed over thousands of years as ground water dissolved limestone. Limestone, which is made of calcium carbonate, dissolves easily in water. As a result, caves form. Some caves reach spectacular proportions, such as the one in **Figure 18.**

Figure 18 *At Carlsbad Caverns, in New Mexico, underground passages and enormous "rooms" have been eroded below the surface of the Earth.*

Cave Formations While caves are formed by erosion, they also show signs of deposition. Water that drips from a crack in a cave's ceiling leaves behind deposits of calcium carbonate. These deposits of calcium carbonate are a type of limestone called *dripstone*. Water and dissolved limestone can drip downward into sharp, icicle-shaped dripstone features known as a stalactites. At the same time, water drops that fall to the cave's floor add to cone-shaped dripstone features known as stalagmites. If water drips long enough, the stalactites and stalagmites can reach each other and join, forming a dripstone column.

Environment CONNECTION

Most bat species live in caves. These night-flying mammals navigate by sound and can reach speeds of 95 km/h. Today scientists know that bats play an extremely important role in the environment. Bats are great consumers of insects, and many bat species pollinate plants and distribute seeds.

IS THAT A FACT!

Mount Shasta is a dormant volcano in the Shasta River basin, in California. During snowmelts, the Shasta River doesn't receive much runoff, however. The snowmelt is absorbed by the porous volcanic slopes and reemerges through springs at the bottom of the river valley.

internetconnect

SCILINKS NSTA

TOPIC: Water Underground
GO TO: www.scilinks.org
***sci*LINKS NUMBER:** HSTE265

3 Extend

GROUP ACTIVITY

MATERIALS

For Each Group:

- 2 jars
- washing soda (sodium carbonate)
- spoon
- dish
- paper clips
- yarn

Tell students to fill the jars with very warm water and stir a few spoonfuls of washing soda into each jar. Students should stop when the soda no longer dissolves. Set the dish between the jars. Clip paper clips to the ends of the yarn, and lower each end into a different jar. After 2 or 3 days, some of the soda-water mixture will have moved through the yarn by capillary action, and the solution will have dripped onto the plate, forming a stalactite and stalagmite by evaporation.

GOING FURTHER

PORTFOLIO Have students research Mammoth Cave, in Kentucky, to learn more about the unique formations there. Then help them construct a "virtual cave." Have students work in groups to create models of the cave's formations using modeling clay. Different groups can model different sections of the cave. Students can also make an audio tape tour of the cave that tells visitors about the structures, including how they formed.

4 Close

Quiz

1. Define *porosity* and *permeability.* (Porosity is the amount of open space between rock particles. Permeability is the ability of rock to allow water to flow through it.)
2. Describe the formation that produces an artesian spring. (A cap rock seals off a permeable layer of rock that is filled with water. If the water table is above the cap rock, water under pressure can escape through an opening in the cap rock.)
3. What is a spring? What is a well? (A spring forms where the water table reaches the surface and water flows out. A well is a human construction that extends below the water table. Water flows into the well's opening and is pumped to the surface.)

Alternative Assessment

Concept Mapping Have students use the following terms to make a concept map:

ground water, water table, aquifer, porosity, permeability, artesian spring, sinkhole, spring, well, cave, dripstone

Sheltered English

Reinforcement Worksheet "Dig It!"

Problem Solving Worksheet "Water Crisis at Happy Acres"

Sinkholes When the water table is lower than the level of a cave, the cave is no longer supported by the water underneath. The roof of the cave can then collapse, leaving a circular depression called a *sinkhole.* Surface streams can "disappear" into sinkholes and then flow through underground caves. Sinkholes often form lakes in areas where the water table is high. Central Florida is covered with hundreds of round sinkhole lakes. **Figure 19** shows how underground caves can affect a landscape.

Figure 19 *This city block shows the effects of a sinkhole in Winter Park, Florida.*

internetconnect

SciLINKS NSTA

TOPIC: Water Underground
GO TO: www.scilinks.org
***sci*LINKS NUMBER:** HSTE265

SECTION REVIEW

1. What is the water table?
2. What is an aquifer?
3. What are some of the features formed by underground erosion and deposition?
4. **Analyzing Relationships** What is the relationship between the zone of aeration, the zone of saturation, and the water table?

18

Answers to Section Review

1. Answers will vary. The water table is the level to which the ground is saturated by water.
2. An aquifer is a rock layer that stores and allows the flow of ground water.
3. Caves and sinkholes form by groundwater erosion. Stalagmites, stalactites, and dripstone columns form by underground deposition.
4. The water table is the underground boundary where the zone of aeration and the zone of saturation meet.

SECTION 4 READING WARM-UP

Terms to Learn

point-source pollution
nonpoint-source pollution
sewage treatment plant
septic tank

What You'll Do

- Describe the stages of treatment for water at a sewage treatment plant.
- Compare a septic system with a sewage treatment plant.
- Explain how ground water can be both a renewable and a nonrenewable resource.

Using Water Wisely

All living things need water to survive. But there is a limited amount of fresh water available on Earth. Only 3 percent of Earth's water is drinkable. And of the 3 percent that is drinkable, 75 percent is frozen in the polar icecaps. That's more than 100 times the volume of water found in the Earth's lakes and streams! This frozen water is not readily available for our use. Therefore, it is important that we use our water resources wisely.

Water Pollution

Surface water, such as rivers and lakes, and ground water are often polluted by waste from cities, factories, and farms. One type of pollution is called **point-source pollution** because it comes from one particular point, such as a sewer pipe or a factory drain. Fortunately, laws prohibit much of this type of pollution.

There is growing concern, however, about another type of pollution, called **nonpoint-source pollution.** This type of pollution, as shown in **Figure 20,** is much more difficult to control because it does not come from a single source. Most nonpoint-source pollution contaminates rivers and lakes by runoff. The main sources of nonpoint-source pollution are street gutters, fertilizers, eroded soils and silt from farming and logging, drainage from mines, and salts from irrigation.

As you know, ground water is an important source of fresh water. In fact, more than half of all household water in the United States comes from ground water. Farms use ground water for irrigation. Because ground water is supplied by water from the Earth's surface, ground water can become contaminated when surface water is polluted. And once polluted, ground water is very difficult to clean up.

Figure 20 *The runoff from this irrigation system could collect pesticides and other pollutants. The result would be nonpoint-source pollution.*

19

IS THAT A FACT!

According to a 1998 U.S. Geological Survey report, water use in the United States has decreased 2 percent since 1990 and almost 10 percent since 1980, despite an increase in population. The report attributes this to public awareness and community conservation efforts.

SECTION 4

Focus

Using Water Wisely

This section discusses water pollution, overuse, and treatment. Students learn about the difference between point-source and nonpoint-source pollution and explore different methods for treating polluted water. The section then discusses trends in domestic, industrial, and agricultural water use and conservation.

Bellringer

Pose the following question on the board or overhead projector:

While hiking, you realize your canteen is almost empty. Why should you not fill it with water from the nearby stream? (Even though the water may look clean, it might contain pollutants or bacteria.)

1 Motivate

DISCUSSION

Ask students to identify all the activities their family does that involve water. Write students' responses on the board. Encourage students to discuss how their lives would be affected if the water supply became depleted or too contaminated to use. Encourage students to track the amount of water they use during the course of a week. Have them use this number to extrapolate how much water they will use in their lifetime.

Directed Reading Worksheet Section 4

2 Teach

Multicultural CONNECTION

Writing Encourage students to use library or Internet resources to investigate traditional American Indian beliefs about the Earth, including its rivers. Students may also be interested to find out about controversies involving American Indian tribes and the use of water resources, such as fisheries. Ask them to write a brief report summarizing their findings.

COOPERATIVE LEARNING

Divide the class into small groups, and challenge each group to create a board game. Tell them that the object of the game is to be the first player to successfully travel through a sewage treatment plant. Provide each group with poster board, plain index cards, and markers. Direct them to create a game board that leads players through the plant and incorporates the concepts they have learned about in this section. Have them use the index cards to write clues and questions directing players' movements through the sewage treatment process. For example, they might write, "If you can define primary treatment, advance to the aeration tank." Have students create written rules, and allow time for the game to be played.

BRAIN FOOD

Each year more than 2.5 million people die from waterborne diseases, such as cholera and typhoid. With better sewage treatment, many of these lives could be saved.

Cleaning Polluted Water

When you flush the toilet or watch water go down the shower drain, do you ever wonder where this water goes? If you live in a city or large town, the water flows through sewer pipes to a sewage treatment plant. **Sewage treatment plants** are factories that clean the waste materials out of water that comes from the sewer or drains. These plants help protect the environment from water pollution. They also protect us from diseases that are easily transmitted through dirty water.

Primary Treatment When water reaches a sewage treatment plant, it is cleaned in two different ways. First it goes through a series of steps known as *primary treatment.* In primary treatment, dirty water is passed through a large screen to catch solid objects, such as paper, rags, and bottle caps. The water is then placed in a large tank, where smaller particles can sink and be filtered out. These particles include things such as food, coffee grounds, and soil. Any floating oils and scum are skimmed off the surface.

Secondary Treatment At this point, the water is ready for *secondary treatment.* In secondary treatment, the water is sent to an aeration tank, where it is mixed with oxygen and bacteria. The bacteria feed on the wastes and use the oxygen. The water is then sent to another settling tank, where chlorine is added to disinfect the water. The water is finally released into a water source—a stream, a lake, or the ocean. **Figure 21** shows the major components of a sewage treatment plant.

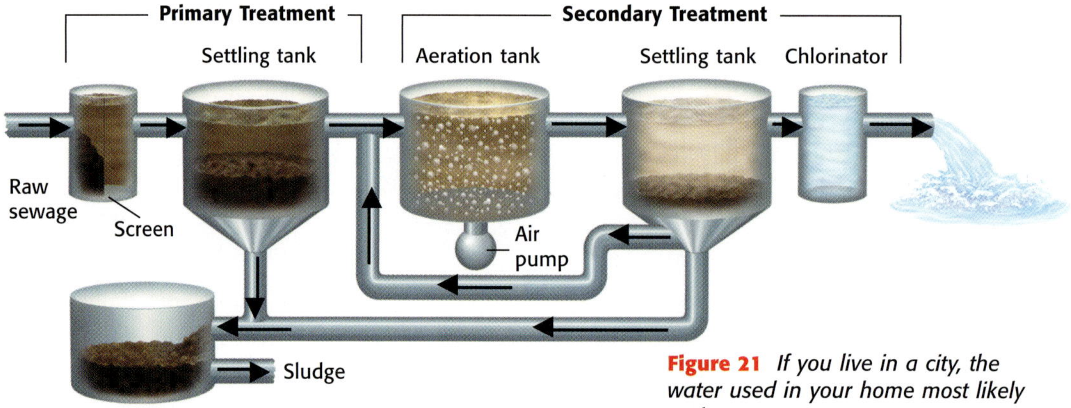

Figure 21 *If you live in a city, the water used in your home most likely ends up at a sewage treatment plant, where it is cleaned by a process that imitates water's natural cleaning cycle.*

Homework

Presentation Have students research sources of nonpoint-source pollution. Ask them to make posters for a community action campaign to reduce nonpoint-source pollution. Display students' posters in the classroom or around school.

MISCONCEPTION ALERT

Students may think that they cannot do anything to conserve water and prevent pollution. Point out to students that there are many things they can do to contribute to conservation. Tell students that by changing a few habits, they can greatly reduce their water consumption and reduce water pollution.

Another Way to Clean Waste Water If you live in an area without a sewage treatment plant, your house probably has a septic tank, such as the one shown in **Figure 22.** A **septic tank** is a large underground tank that collects and cleans waste water from a household. Waste water flows from the house into the tank, where the solids sink to the bottom. Bacteria consume these wastes on the bottom of the tank. The water flows from the tank into a group of buried pipes. The buried pipes distribute the water, enabling it to soak into the ground. This group of pipes is called a *drain field.*

LabBook

Get your hands dirty and learn about some of the methods used to clean up water. Check out page 100 of the LabBook.

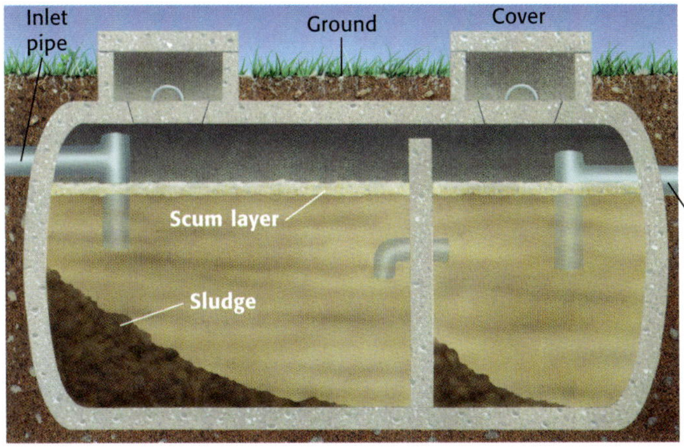

Figure 22 *Most septic tanks must be cleaned out every few years in order to work properly.*

Where the Water Goes

The chart in **Figure 23** shows how an average household in the United States uses water. Notice that less than 8 percent of the water we use in our homes is used for drinking. The rest is used for flushing toilets, doing laundry, bathing, and watering lawns and plants.

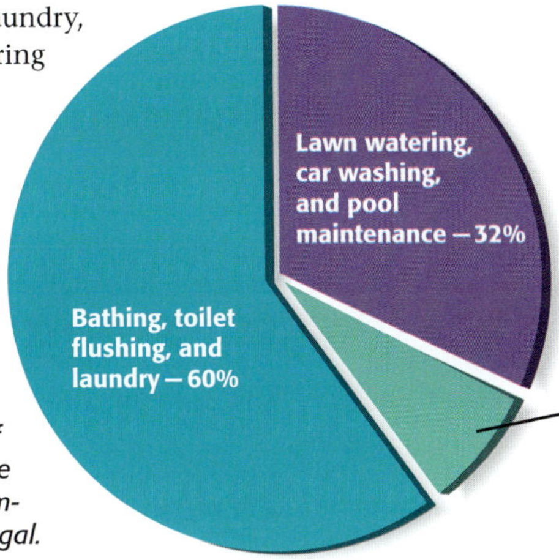

Figure 23 *The average household in the United States uses about 100 gal of water per day. This pie chart shows some common uses of this 100 gal.*

Activity

Study the chart at left and determine where the majority of water is used. Think of some ways that you can decrease the amount of water that you use in your home. Share your suggestions with your class.

TRY at HOME

LabBook PG 100
Clean Up Your Act

MATH and MORE

Low-flush toilets use about 6 L of water per flush. Standard toilets use 19 L of water per flush. How much water is saved in a week by replacing a standard toilet with a low-flush toilet? Assume the toilet is flushed 10 times a day. (910 L)

How much water would be saved in a year? (47,450 L)

Math Skills Worksheet "Multiplying Whole Numbers"

DEBATE

Should Water Conservation Be Enforced? Point out to students that many communities are enacting mandatory water conservation measures. For example, Albuquerque, New Mexico, has adopted a water-conservation policy in an effort to reduce per capita water use by 30 percent. However, some people are opposed to government regulation of water use. Divide the class into two groups, and assign each group a position in the debate. Have students use library or Internet resources to research their position.

IS THAT A FACT!

Mining can be a major source of water pollution. Although coal and mineral ores are part of Earth's rock formations and soil, some can be dangerous if they are concentrated in ground water and rivers and streams. Four major river basins in Pennsylvania are being polluted by runoff from thousands of abandoned coal mines. As a result, more than 4,800 km of stream water and ground water have been contaminated with acidic runoff containing toxic metals. The impact has been enormous; many of the streams are now devoid of fish.

Answers to Activity

Sample answers: I will make sure the dishwasher is completely full when it is being used. If I'm washing dishes in the sink, I won't let the water run continuously. I'll turn the shower off when lathering up. And I won't do several small loads of laundry if I can do fewer large loads instead.

3 Extend

CONNECT TO ENVIRONMENTAL SCIENCE

Settlement of the American West followed the conquest of its rivers. Damming and controlling rivers created usable farmland, allowed settlement in flood plains, and powered industry. Beginning in the 1930s, massive hydroelectric dams were built across many western rivers. Following this example, many countries have begun to dam their rivers to provide water for irrigation and drinking, to control flooding, and to produce electricity. While dams provide many benefits, large dam projects face increasing criticism for numerous social, environmental, economic, and safety concerns. Around the world, as many as 60 million people have been displaced by dams. Even more have suffered from the effects of dams downstream: farmland deprived of flood sediments and water for irrigation becomes unusable, fisheries become less productive, and epidemics of waterborne disease often follow large dam projects. Although hydropower is hailed as a cheap, nonpolluting source of energy, the true cost and impact of hydropower is often great. Should projects such as China's Three Gorges Dam be stopped? After America has benefitted so much from dam projects, can we ask developing countries not to exploit their water resources? Encourage students to find out more and debate these issues in class.

Answer to Apply

Answers will vary. Accept all reasonable responses.

Water in Industry The chart on the previous page shows how fresh water is used in homes. Even more water is required for industry, as shown in **Figure 24.** Water is used to cool power stations, to clean industrial products, to extract minerals, and to create power for factories. Many industries are trying to conserve water by reusing it in their production processes. In the United States, most of the water used in factories is recycled at least once. At least 90 percent of this water can be treated and returned to surface water.

Ground-water supplies also need to be monitored. Although ground water is considered to be a *renewable resource,* a resource that can be replenished, recycling ground water can be a lengthy process. When overused, ground water can sometimes be categorized as a *nonrenewable resource,* a substance that cannot be replaced once it is used. Ground water collects and moves slowly, and water taken from some aquifers might not be replenished for many years.

Figure 24 *The core of a nuclear reactor is cooled by water.*

APPLY

Wasting Water?

How much water do you use when you brush your teeth? Picture yourself at home brushing your teeth. Time how long it takes you to go through the procedure. In your ScienceLog, write down the steps you take, making sure to include how many times you turn on and turn off the faucet. During what percentage of the time spent brushing your teeth is the water running? How do you think you might be wasting water? What are some ways that you could conserve water while brushing your teeth?

22

internetconnect

SciLINKS NSTA

TOPIC: Water Pollution and Conservation
GO TO: www.scilinks.org
***sci*LINKS NUMBER:** HSTE270

Water in Agriculture The Ogallala aquifer is the largest known aquifer in North America. The map in **Figure 25** shows that the Ogallala aquifer runs beneath the ground through eight states, from South Dakota to Texas. For the last 100 years, the aquifer has been used heavily for farming. The Ogallala aquifer provides water for approximately one-fifth of the crop-land in the United States. Recently, the water table in the aquifer has dropped so low that some scientists say that it would take at least 1,000 years to replenish the aquifer if it were no longer used.

BRAIN FOOD

The Ogallala aquifer can hold enough water to fill Lake Huron. At this time, however, the aquifer is being used 25 times as fast as it is being replenished.

Figure 25 *Because the Ogallala aquifer has been such a good source of ground water, it has become overused. The water table has dropped more than 30 m in some areas.*

Water resources are different from other resources. Because water is necessary for life, there is no alternative resource. Aquifers are often overused and therefore do not have time to replenish themselves. Like surface water, ground water must be conserved.

SECTION REVIEW

1. What is the difference between nonpoint-source and point-source pollution?
2. Summarize the process of water treatment in a sewage treatment plant.
3. What is the difference between a renewable resource and a nonrenewable resource?
4. **Summarizing Data** How does a septic tank work?

internet connect

SciLINKS NSTA

TOPIC: Water Pollution and Conservation
GO TO: www.scilinks.org
***sci*LINKS NUMBER:** HSTE270

23

Answers to Section Review

1. Point-source pollution comes from a single, identifiable source, whereas nonpoint-source pollution does not.
2. When water enters a sewage plant, it goes through two treatments. In primary treatment, most of the solid material is settled out. In secondary treatment, the water is cleaned by bacteria that feed on the waste and chlorine that kill the bacteria.
3. A renewable resource can be replaced or recycled. Once a nonrenewable resource is used up, it cannot be replaced.
4. Used water from a home flows into an underground tank, where bacteria feed on the solid waste. The rest of the water flows through underground pipes, which distribute it into the surrounding soil.

4 Close

Quiz

1. What are two sources of nonpoint-source pollution? (Sample answers: street gutters, fertilizers, eroded soils and silt from farming and logging, drainage from mines, and salts from irrigation)
2. How do sewage treatment plants protect the environment? (They clean polluted water before it is returned to a water source.)

ALTERNATIVE ASSESSMENT

PORTFOLIO

Provide students with poster board and markers. Direct them to create posters illustrating how nonpoint-source pollution can cause contamination of both surface water and ground water. Their diagrams should reflect the understanding that pollution of surface water can spread to ground water. Sheltered English

COOPERATIVE LEARNING

Divide the class into small groups. Challenge each group to write a public-service announcement to educate the public about the need to conserve water and to avoid polluting ground water. The announcements should include definitions of nonrenewable and renewable resources and outline practical steps that everyone can take to protect water resources. Sheltered English

Interactive Explorations CD-ROM "Flood Bank"

Water Cycle—What Goes Up . . .
Teacher's Notes

Time Required
One 45-minute class period

Lab Ratings
EASY → HARD

TEACHER PREP: 1

STUDENT SET-UP: 2

CONCEPT LEVEL: 2

CLEAN UP: 1

MATERIALS
The materials listed on the student page are enough for a group of 4–5 students.

Safety Caution
Remind students to review all safety cautions and icons before beginning this lab activity. Students should be cautioned when using a hot plate. Care should also be exercised in using the glassware. Appropriate methods should be used to dispose of broken glass.

Norman Holcomb
Marion Local Schools
Maria Stein, Ohio

Making Models Lab

Water Cycle—What Goes Up . . .

Why does a bathroom mirror fog up? Where does water go when it dries up? Where does rain come from, and why don't clouds run out of rain? These questions relate to the major parts of the water cycle—condensation, evaporation, and precipitation. In this activity, you will make a model of the water cycle, and you will watch water as it moves through the model.

MATERIALS

- graduated cylinder
- 50 mL of tap water
- heat-resistant gloves
- beaker
- hot plate
- glass plate or watch glass
- tongs or forceps

Procedure

1. Use the graduated cylinder to pour 50 mL of water into the beaker. Note the water level in the beaker.
2. Put on your safety goggles and gloves. Place the beaker securely on the hot plate. Turn the heat to medium, and bring the water to a boil.
3. While waiting for the water to boil, practice picking up and handling the glass plate or watch glass with the tongs. Hold the glass plate a few centimeters above the beaker, and tilt it so that the lowest edge of the glass is still above the beaker.
4. Observe the glass plate as the water in the beaker boils. In your ScienceLog, write down the changes you see in the beaker, in the air above the beaker, and on the glass plate. Write down any changes you see in the water.
5. Continue until you have observed steam rising off the water, the glass plate becoming foggy, and water dripping from the glass plate.
6. Carefully set the glass plate on a counter or other safe surface as directed by your teacher.
7. Turn off the hot plate, and allow the beaker to cool. If you are directed to do so by your teacher, move the hot beaker with gloves or tongs.
8. In your ScienceLog, copy the illustration shown on the next page. On your sketch, draw and label the water cycle you observed in your model. Include arrows and labels for condensation, evaporation, and precipitation.

24

Analysis

9 Compare the water level in the beaker now with the water level at the beginning of the experiment. Was there a change? Explain why or why not.

10 If you had used a scale or a balance to measure the mass of the water in the beaker before and after this activity, would the mass have changed? Why or why not?

11 How is your model similar to the Earth's water cycle? On your sketch of the illustration above, label where the processes shown in the model are similar to the Earth's water cycle.

12 When you finished this experiment, the water in the beaker was still hot. What stores much of the energy in the Earth's water cycle?

Going Further

As rainwater runs over the land, the water picks up minerals and salts. Do these minerals and salts evaporate, condense, and precipitate as part of the water cycle? Where do they go?

If the average global temperature on Earth becomes warmer, how would you expect sea levels to change, and why? What if the average global temperature cools?

Making Models Lab

Answers

8. Answer is at the bottom of the page.

9. The water level is less at the end of the experiment than at the beginning of the experiment because some of the water escaped in the form of steam.

10. The mass would have changed slightly. Because some of the steam escaped, the mass of the water in the beaker would be reduced.

11. The model is similar to the Earth's water cycle because precipitation, condensation, and evaporation occurred. Accept all reasonable depictions of the water cycle.

12. Much of the energy in the water cycle is stored in bodies of water such as the oceans. Energy is also stored in the atmosphere and the Earth's land surface.

Going Further

- No; minerals and salts are not part of the water cycle. As water evaporates, minerals and salts are left behind as deposits.
- If the average global temperature gets warmer, the sea level may rise because the polar icecaps would melt. If the average global temperature gets cooler, the sea level may decrease as more water is frozen.

8. Student sketches should resemble the water cycle illustrated below:

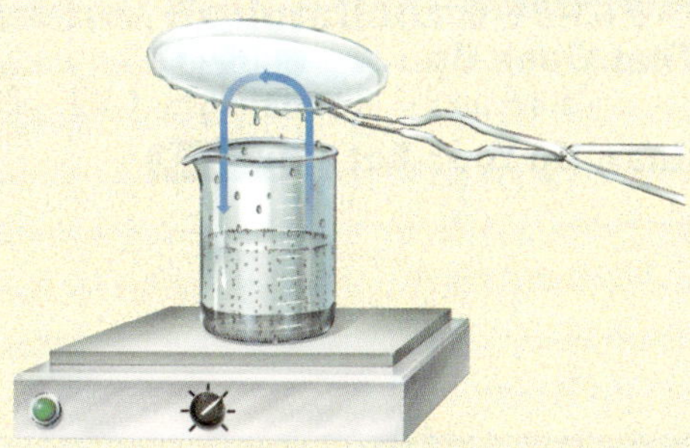

Datasheets for LabBook

Science Skills Worksheet
"Science Drawing"

Weird Science
Bubble, Boil, & Squirt

Background

Yellowstone National Park, located in northwestern Wyoming and southern Montana, includes more than 10,000 geothermal features. All of these features are the product of Yellowstone's unique geographic position. About half of Yellowstone Park lies on top of three major calderas that formed between 0.6 million and 2 million years ago. A caldera is a circular depression formed when the roof of a magma chamber collapses as a result of a volcanic eruption. The heat from the most recent caldera accounts for the large number of geothermal features in the park.

Fumaroles are the vents that emit the volcanic gas responsible for forming mud pots.

Weird Science

Bubble, Boil, & Squirt

In parts of Yellowstone National Park boiling water blasts into the sky, lakes of strange-colored mud boil and gurgle, and hot gases hiss from the ground. What are these strange geologic features? What causes them? The story begins deep in the Earth.

Old Geysers

One of Yellowstone's main tourist attractions is a *geyser* called Old Faithful. Erupting every 60 to 70 minutes, Old Faithful sends a plume of steam and scalding-hot water as high as 60 m into the air. A geyser is formed when a narrow vent connects one or more underground chambers to Earth's surface. These underground chambers are heated by nearly molten rock. As underground water flows into the vent and chambers, it is heated above 100°C. The superheated water quickly turns to steam and explodes first toward the surface and then into the air. And Old Faithful erupts right on schedule!

Nature's Hot Tub

A *hot spring* is a geyser without pressure. Its vents are wider than a geyser's, and they let the underground water cool a little and flow to the surface rather than erupt in a big fountain. To be called a hot spring, the water must be at least as warm as human body temperature (37°C). Some underground springs are several hundred degrees Celsius.

Flying Mud Pies

Mud pots form when steam or hot underground water trickles to the surface and chemically weathers and dissolves surface features, such as rocks. The mixture of dissolved rock and water creates a boiling, bubbling pool of sticky liquid clay. But don't get too close! Occasionally, the steam will rise quickly and forcefully enough to make the mud pot behave like a volcano. When it does, a mud pot can toss car-sized gobs of mud high into the air!

Some mud pots become *paint pots* when microorganisms or brightly colored minerals are mixed in. For instance, if there is a lot of iron in the mud, the paint pot will turn reddish brown or yellowish brown. Other minerals and bacteria can make the mud white or bluish in color. Some paint pots may even gurgle up blobs in several different colors.

▲ *Mud Pot in Yellowstone National Park*

What Do You Think?

▶ Some people believe that tapping geothermal energy sources such as geysers could harm the delicate ecology of those sources. Find out about the benefits and the risks of using geothermal energy. What is your opinion?

30

Sample Answer to What Do You Think?

The main advantage of geothermal energy is that, unlike fossil fuels, it doesn't produce pollution and is a renewable resource. Also, geothermal energy plants require less space than normal power plants. However, some scientists believe that using geothermal energy from geysers could disrupt the natural flow of hot springs and possibly disrupt ground-water sources. Some geysers recently used to make electricity have stopped producing steam. Other possible consequences of the disruption of geothermal areas are as follows: the extinction of microorganisms that thrive in hot waters, collapsing calderas, and more-frequent volcanic eruptions.

EYE ON THE ENVIRONMENT

Disaster Along the Delta

As the sun rises over the delta wetlands of the Mississippi River, fishermen test their skills. Long-legged birds step lightly through the marsh, hunting fish or frogs for breakfast. And hundreds of species of plants and animals start another day in this fragile ecosystem. But the delta ecosystem is in danger of being destroyed.

The threat comes from efforts to make the Mississippi more useful. Large portions of the river bottom were dredged to make the river deeper for ship traffic. Underwater channels were built to control flooding. What no one realized was that sediments that were once deposited to form new land now pass through the deep channels and flow out into the ocean.

Those river sediments replaced the land that was lost every year to erosion. Without the sediments, the river can't replace the land lost to erosion. And so the Mississippi River delta is disappearing. By 1995, more than half the wetlands were already gone, swept out to sea by waves along the Louisiana coast.

▲ *The Mississippi River flows from Minnesota through the Midwest to the Gulf of Mexico in the southern United States.*

Sedimental Journey

The Mississippi River journeys 3,766 km to empty 232 million metric tons of sediment into the Gulf of Mexico each year. The end of the Mississippi River delta forms the largest area of wetlands in North America. A *delta* forms when sediments settle at the mouth of a river. At the Mississippi River delta, the sediments build up and form new land along the Louisiana coastline. The area around the delta is called *wetlands.* It has fertile soil, which produces many crops, and a variety of habitats—marsh, freshwater, and saltwater—that support many species of plants and animals.

Taking Action to Preserve the Delta

Since the mid-1980s, local, state, and federal governments, along with Louisiana citizens and businesses, have been working together to monitor and restore the Mississippi River delta. Some projects to protect the delta include filling in canals that divert the sediments and even using old Christmas trees as fences to trap the sediments! With the continued efforts of scientists, government leaders, and concerned citizens, the Mississippi River delta stands a good chance of recovering.

Explore the Delta

► Find out more about the industries and organisms that depend on the Mississippi River delta for survival. What will happen to them if we don't take care of the ecosystem?

31

Answers to Explore the Delta

The coastal wetlands at the Mississippi River delta provide a habitat for diverse communities of plants and animals, such as fish, shellfish, and birds. The wetlands are the nursery and feeding area for millions of waterfowl. These species and the industries that rely on them, such as fishing and tourism, will suffer greatly if the wetland ecosystems are not protected. If erosion is severe enough, buildings, highways, phone lines, and pipelines will need to be relocated.

EYE ON THE ENVIRONMENT

Disaster Along the Delta

Background

The word *Mississippi* comes from a Native American word that means "big river." The Mississippi River flows for 3,766 km from its source, in Minnesota, to its mouth, in the Gulf of Mexico. More than 2,897 km of the river can be navigated by ships and boats for commercial, transportational, and recreational uses.

The Mississippi River forms part of the boundaries of 10 different states. As it travels to the Gulf of Mexico, other rivers—such as the Ohio and the Missouri—flow into it and increase the volume of water in the Mississippi River.

The Mississippi River delta covers about 33,700 km^2. It continuously nourishes the coastal wetlands that provide a habitat for diverse populations of plants and animals.

The erosion and sinking of land near the Mississippi River delta has seriously disrupted local economies. For instance, the coastal commercial and recreational fishing industries contribute about $2 billion to Louisiana's economy every year and provide about 50,000 to 70,000 jobs for the people of Louisiana. Also, the coastal wetlands provide coastal residents with protection from hurricanes and other storms. Currently, the erosion in this area accounts for 80 percent of the nation's loss of coastal wetlands per year.

Chapter Resources & Worksheets

Visual Resources

TEACHING TRANSPARENCIES

#149

#150

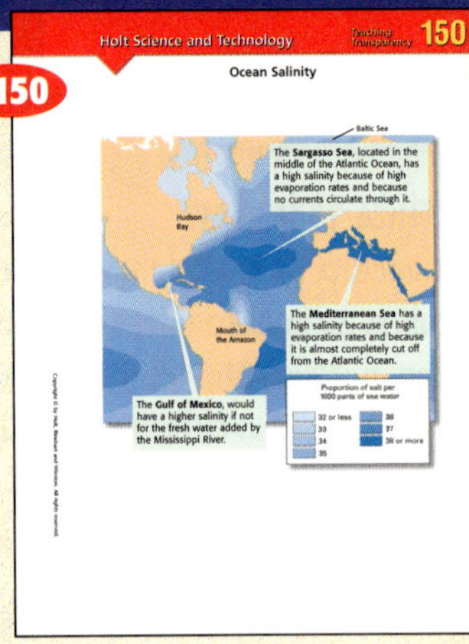

#151

#152

#153

CONCEPT MAPPING TRANSPARENCY

TEACHING TRANSPARENCIES

#154

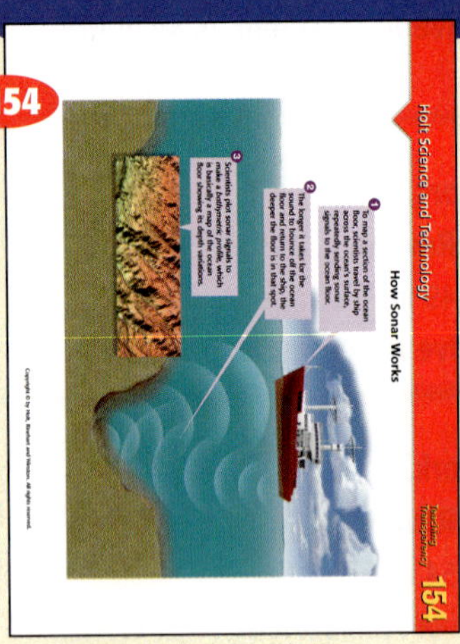

#155

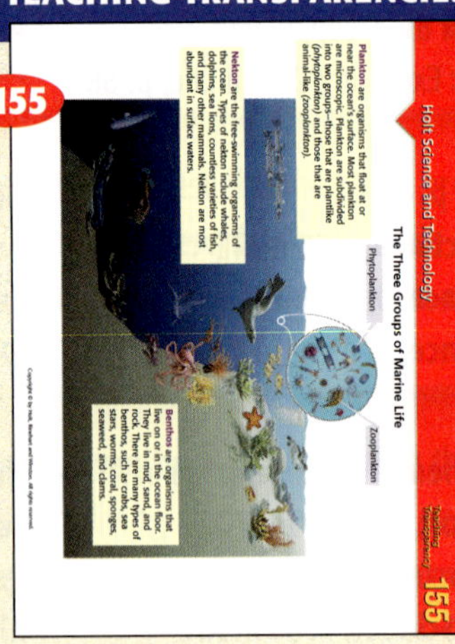

#27

#13

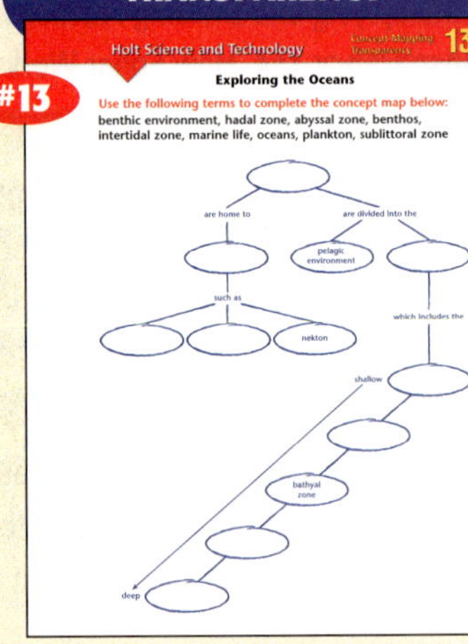

Meeting Individual Needs

DIRECTED READING

#2

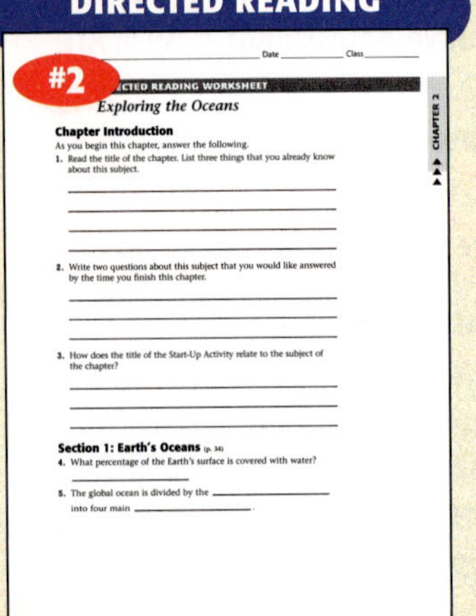

REINFORCEMENT & VOCABULARY REVIEW

#2

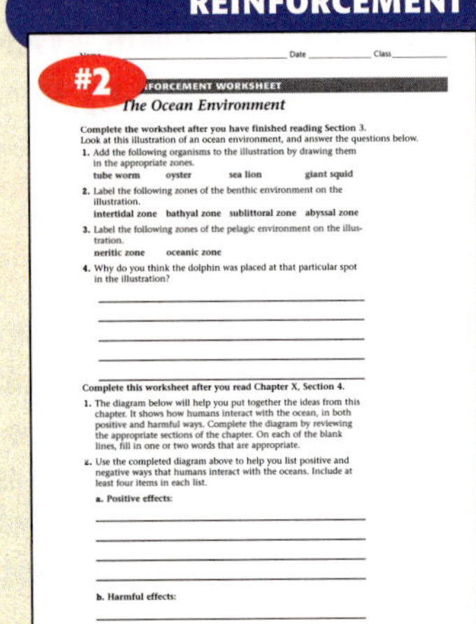

#2

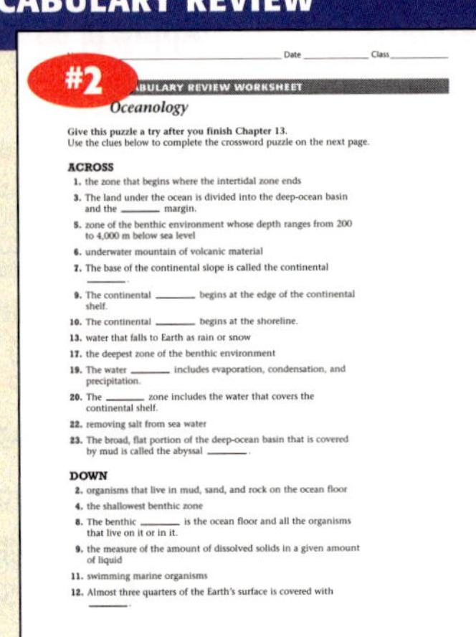

SCIENCE PUZZLERS, TWISTERS & TEASERS

#2

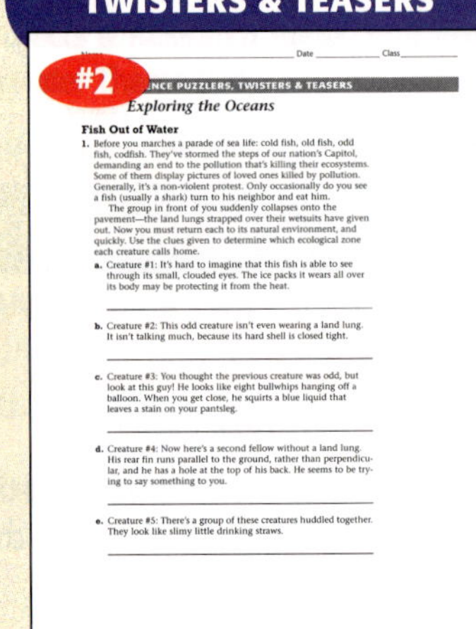

Chapter 2 • Exploring the Oceans

Review & Assessment

STUDY GUIDE

#2 Vocabulary & Notes Worksheet — *Exploring the Oceans*

#2 Chapter Review Worksheet — *Exploring the Oceans*

CHAPTER TESTS WITH PERFORMANCE-BASED ASSESSMENT

#2 Exploring the Oceans — *Chapter 2 Test*

#2 Exploring the Oceans — *Chapter 2 Performance-Based Assessment* (Skill Builder)

Lab Worksheets

INQUIRY LABS

#2 Student Worksheet — *Surf's Up!* (Making Models)

LONG-TERM PROJECTS & RESEARCH IDEAS

#2 Student Worksheet — *Your Very Own Underwater Theme Park* (Design Your Own)

ECOLABS & FIELD ACTIVITIES

#2 Student Worksheet — *Operation Oil-Spill Cleanup* (Discovery Lab)

WHIZ-BANG DEMONSTRATIONS

#2 Teacher-Led Demonstration — *Fowl Play* (Discovery Lab)

DATASHEETS FOR LABBOOK

#2 Probing the Depths

#2 Investigating an Oil Spill

Applications & Extensions

CRITICAL THINKING & PROBLEM SOLVING

#2 Critical Thinking Worksheet — *Chain Reaction*

MULTICULTURAL CONNECTIONS

#9 Science in the News: Critical Thinking Worksheets — Segment 9 — Saving Pacific Sea Horses

SCIENTISTS IN ACTION

#17 Science in the News: Critical Thinking Worksheets — Segment 17 — Deep Flight

#18

INTERACTIVE EXPLORATIONS

#2-2 Exploration 2 Worksheet — Sea Sick

Chapter Background

Section 1

Earth's Oceans

▶ The Global Ocean

Historically, the global ocean was divided into five oceans: the Atlantic, Pacific, Indian, Arctic, and Antarctic Oceans. Today, most oceanographers agree that the Antarctic Ocean is actually the southernmost section of the Atlantic, Pacific, and Indian Oceans.

▶ The Mariana Trench

In 1960, the *Trieste,* a small submersible designed to explore the ocean to great depths, set out on a voyage that until then had only been imagined: it descended into the Mariana Trench, the deepest known place on the Earth. As Jacques Piccard and a companion descended, they were surprised to feel abrupt changes between the ocean's temperature layers. Every time the vessel reached the boundary between two layers, it seemed to stop as though it had reached the ocean floor.

IS THAT A FACT!

- The average depth of each ocean is as follows:

  Arctic: 1,038 m
  Atlantic: 3,735 m
  Indian: 3,872 m
  Pacific: 4,188 m

- The average depth for all the oceans is about 3,800 m.

Section 2

The Ocean Floor

▶ The Renewal of a Planet

Submersible missions such as *Alvin*'s have enabled oceanographers to witness the creation of oceanic crust and the forces that drive tectonic plate movement. By observing molten rock welling up into the ocean, they have seen how new seafloor forms and have gained a better understanding of how the continents drift apart. Trained to "read" rock formations, these scientists use their observations to reconstruct the Earth's history.

IS THAT A FACT!

- The oceans' deep-water sound channels carry sound waves for hundreds of kilometers. Whales and other marine animals take advantage of these properties for long-range communication and to search for food. Whales communicate with clicks, whistles, squeaks, and songs that convey information. Scientists aren't sure what these songs mean, but they do know that whales can communicate at distances as great as 1,600 km!

Section 3

Life in the Ocean

▶ A World of Its Own

In 1977, scientists aboard *Alvin* witnessed an astonishing new world around deep-sea vents. Exploring hydrothermal vents 320 km off the Galápagos Islands, these explorers saw an amazing multitude of unusual marine populations—giant clams and worms, fish, and crabs—gathered in an abyssal oasis.

- Using a special claw attachment, scientists harvested samples of the marine life they found. When they analyzed their specimens, they were in for a surprise—the water smelled like rotten eggs! This smell came from hydrogen sulfide dissolved in the water around the vents.

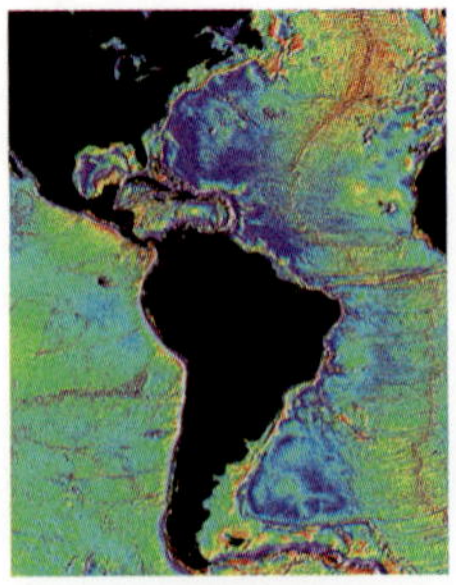

- The scientists discovered that certain marine bacteria thrive on the hydrogen sulfide released by vents in the ocean floor. These bacteria are food for the other marine creatures, and they are part of a food chain that does not rely on photosynthesis for energy.

Creatures of the Dark

The most imaginative science-fiction author would be hard pressed to conceive of creatures as bizarre as the organisms inhabiting the deep sea. Living in perpetual darkness, some—such as the cookie-cutter shark—are bioluminescent. Others lack any organs or structures for vision or photosensitivity. Many benthic (ocean bottom) animals look more like plants than animals. Some exhibit such unique methods of feeding, reproduction, and movement that they are difficult to classify under our current system.

IS THAT A FACT!

- The word *plankton* comes from the Greek *planktos,* meaning "wandering." Because plankton float at or near the ocean's surface, they "wander" with the ocean currents. Interestingly, tiny plankton are the sole sustenance of two of the largest marine animals—the blue whale and the basking shark. The word *planets* is also derived from the same root and refers to the observation that the planets appear to wander among the stars.

SECTION 4

Resources from the Ocean

Food for Thought

Fish are an invaluable ocean resource, providing a significant percentage of the world's protein needs every year. About 75 million tons of fish are harvested from the ocean each year. But many fish populations have been depleted by overfishing.

- Concerned that too few fish will remain to breed, scientists determine the maximum sustainable yield, or the amount of fish that can be harvested each year without jeopardizing future catches. Using the scientists' guidelines, governments sometimes impose fishing restrictions to manage fish populations. Many people are working to ensure that the world can continue to count on fish for food.

Sea Thermal Plants

Harnessing tidal and wave energy is not the only way to get electrical power from the ocean. Since 1979, the United States government has operated an Ocean Thermal Energy Conversion (OTEC) plant off the coast of Hawaii, where the temperature differential between surface and deeper water layers is converted into electrical energy.

SECTION 5

Ocean Pollution

Thermal Pollution

Pesticides, oil, sludge, and trash are not the only harmful pollutants released into our oceans. Power plants can cause thermal pollution by releasing heated water into the sea. Thermal pollution may result in only a 1- or 2-degree temperature difference in the area near the heat source, but that can have profound effects on the ecology. Fish populations may migrate away from the affected area, and overgrowth of other organisms, or "algal blooms," may occur.

Close Quarters

Closed-in seas are particularly vulnerable to damage from ocean pollution. The shores and adjacent waters of the Mediterranean, Baltic, and Adriatic Seas have been fouled with city sewage, factory waste, and fertilizer and pesticide runoff from farms. Their open waters have been affected by dumping and oil spills.

For background information about teaching strategies and issues, refer to the ***Professional Reference for Teachers.***

CHAPTER 2

Exploring the Oceans

Pre-Reading Questions

Students may not know the answers to these questions before reading the chapter, so accept any reasonable response.

Suggested Answers

1. The oceans have changed in size, shape, and salinity over time. Other changes that are not covered in this chapter include changes in marine life, water chemistry, and sea level.
2. Technologies that scientists use to study the ocean without going under water include sonar, satellites, seismic technology, and dredging.
3. Answers will vary. Some resources mentioned in this chapter include fish, oil, natural gas, fresh water, minerals, tidal energy, and wave energy.

CHAPTER 2

Exploring the Oceans

Sections

Pre-Reading Questions

1. How have Earth's oceans changed over time?
2. Name two ways to study the ocean without going under water.
3. Name two valuable resources that are taken from the ocean.

32

internetconnect

HRW On-line Resources
go.hrw.com
For worksheets and other teaching aids, visit the HRW Web site and type in the keyword: **HSTOCE**

www.scilinks.com
Use the *sci*LINKS numbers at the end of each chapter for additional resources on the **NSTA** Web site.

www.si.edu/hrw
Visit the Smithsonian Institution Web site for related on-line resources.

www.cnnfyi.com
Visit the CNN Web site for current events coverage and classroom resources.

A New World Under Water

Seventy-one percent of the Earth's surface is covered by ocean water. However, large portions of the oceans, especially the deepest parts and the parts nearest the poles, remain completely unexplored. In the past several decades, new technologies have made underwater exploration possible, allowing scientists to gather important information about the Earth's greatest resource. In this chapter, you'll learn about the fascinating world under the water and the important part the oceans play in making our planet livable.

START-UP Activity

EXIT ONLY?

To study what life under water would be like, scientists sometimes live in underwater laboratories. How do these scientists enter and leave these labs? Believe it or not, the simplest way is through a hole in the lab's floor. You might think water would come in through the hole, but it doesn't. People inside the lab can breathe freely and can come and go through the hole at any time. How is this possible? Do the following activity to find out.

Procedure

1. Fill a **large bowl** about two-thirds full of **water.**
2. Turn a **clear plastic cup** upside down.
3. Slowly guide the cup straight down into the water. Be careful not to tip the cup.
4. Record your observations in your ScienceLog.

Analysis

5. How does the air inside the cup affect the water below the cup?
6. How do your findings relate to the hole in the bottom of the underwater research lab?

33

START-UP Activity

EXIT ONLY?

MATERIALS
For Each Group: • large bowl • small, clear plastic cup

Teacher's Notes

A cylindrical cup (one that does not taper) is best for this activity.

Answers to START-UP Activity

5. Sample answer: The air inside the cup prevented the water below it from filling the space inside the cup.
6. Answers will vary. Sample answer: Just as the air in the cup keeps the water from filling the cup, the air in the underwater research lab keeps water from coming through the hole in the bottom of the lab.

SECTION 1

Focus

Earth's Oceans

In this section students learn how the global ocean is divided as well as how the oceans formed. They explore the properties of ocean water, including factors that affect salinity, temperature zones, and surface temperature changes. Finally, they learn how the ocean interacts with the atmosphere and the land via the water cycle.

Bellringer

Show students a photo of Earth from outer space, and ask them to list some of the planet's most obvious features. Have them predict the percentage of land and water on Earth. Tell students that water covers 71 percent of Earth's surface and that liquid water is very rare in our solar system. Discuss the significance of Earth's abundant water resources.

1 Motivate

ACTIVITY

Discuss the divisions of the global ocean shown on this page, noting that the volume of the Pacific Ocean is 724 million cubic kilometers. The volume of the Atlantic Ocean is about 322 million cubic kilometers, the volume of the Indian Ocean is about 292 million cubic kilometers, and the volume of the Arctic Ocean is about 12 million cubic kilometers. Have students use graduated cylinders filled with water to demonstrate these ratios. (The Pacific Ocean would be 724 mL, the Atlantic Ocean would be 322 mL, the Indian Ocean would be 292 mL, and the Arctic Ocean would be 12 mL.)

SECTION 1 READING WARM-UP

Terms to Learn

salinity
thermocline
water cycle

What You'll Do

- Name the major divisions of the global ocean.
- Describe the history of Earth's oceans.
- Summarize the properties and other aspects of ocean water.
- Summarize the interaction between the ocean and the atmosphere.

Earth's Oceans

Earth stands out from the other planets in our solar system primarily for one reason—71 percent of the Earth's surface is covered with water. Most of Earth's water is found in the global ocean, which is divided by the continents into four main oceans. This is shown in the figure below. The ocean is a unique body of water that plays many roles in regulating Earth's environment. Read on to learn more about one of our most important resources—the ocean.

Divisions of the Global Ocean

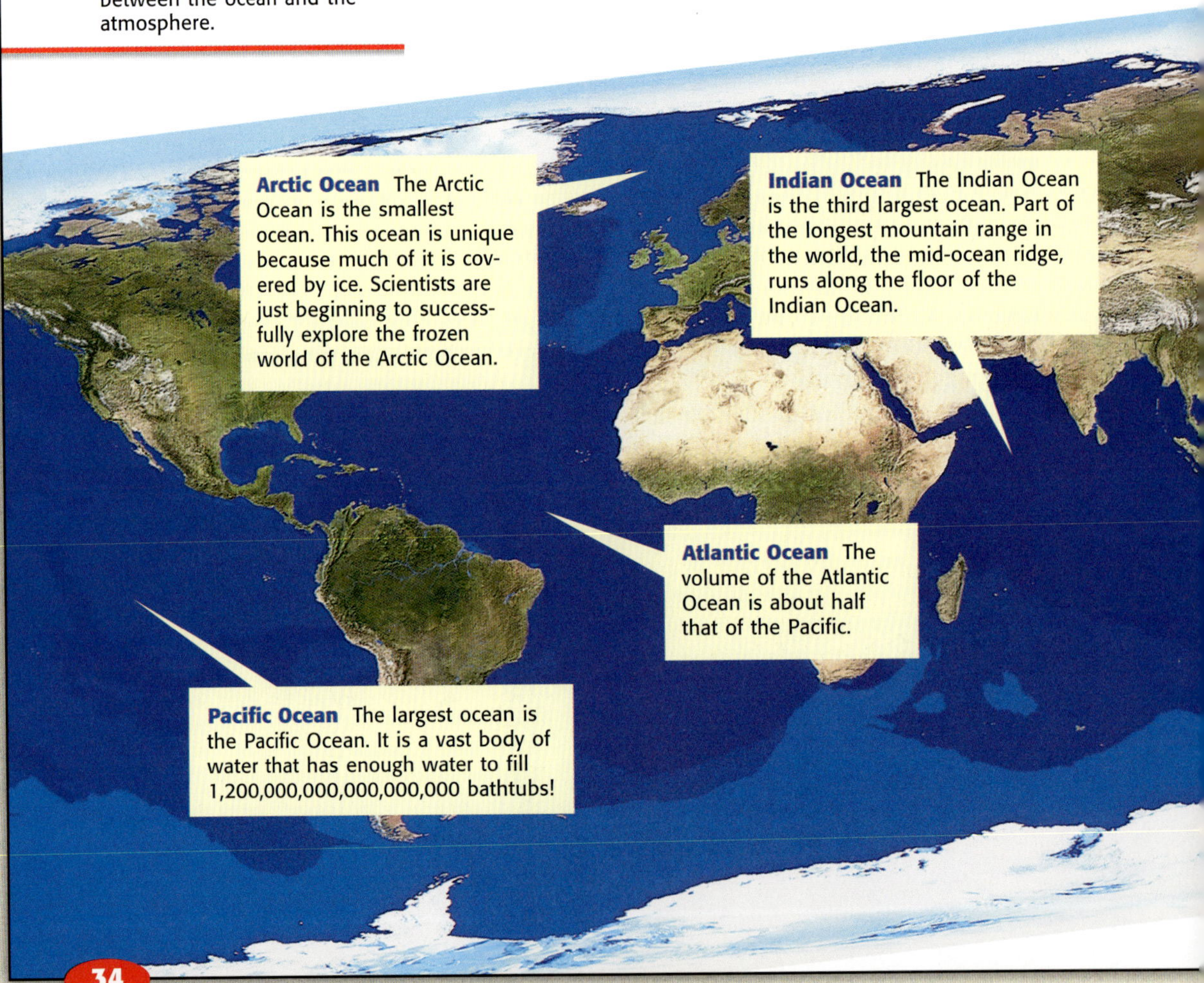

Teaching Transparency 149 "Divisions of the Global Oceans"

Directed Reading Worksheet Section 1

IS THAT A FACT!

The global ocean covers nearly 376 million square kilometers. The entire North American continent, by comparison, covers only a little more than 24 million square kilometers.

How Did the Oceans Form?

About four and a half billion years ago, the Earth was a very different place. There were no oceans. Volcanoes spewed lava, ash, and gases all over the planet, which was much hotter than it is today. The volcanic gases, including water vapor, began to form Earth's atmosphere. While the atmosphere developed, the Earth was cooling. Sometime before 4 billion years ago, the Earth cooled enough for water vapor to condense and fall as rain. The rain began filling the lower levels of Earth's surface, and the first oceans began to form.

Earth's oceans have changed a lot throughout history. Scientists who study oceans have learned much about the oceans' history, as shown in the diagram below.

Self-Check

Examine the diagram below. If North America and South America continue to drift westward and Asia continues to drift eastward, what will eventually happen? *(See page 136 to check your answer.)*

The Recent History of Earth's Oceans

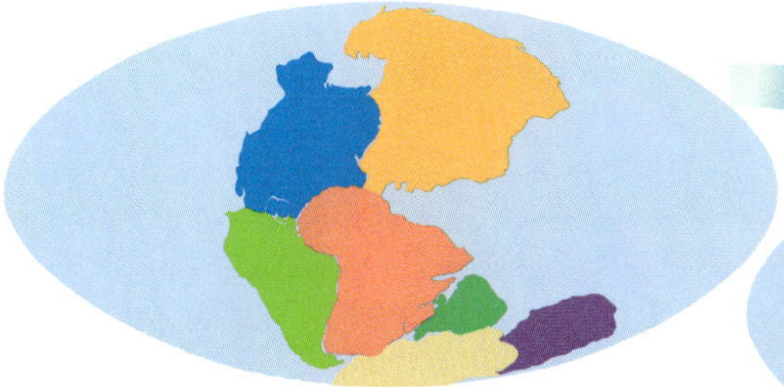

About 245 million years ago, the continents were one giant landmass called *Pangaea* and the oceans were one giant water body called *Panthalassa.*

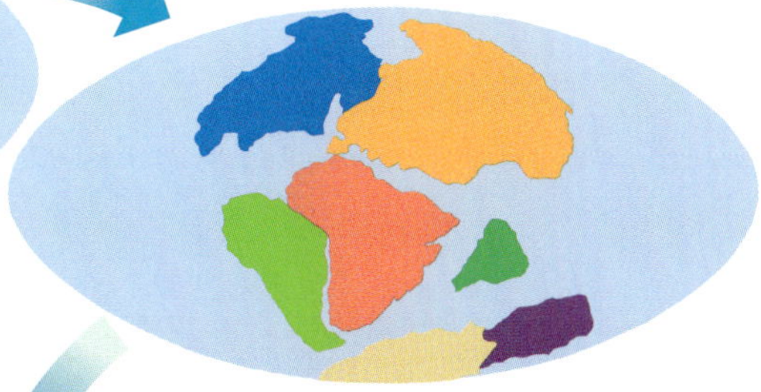

About 180 million years ago, the North Atlantic Ocean and the Indian Ocean began to form.

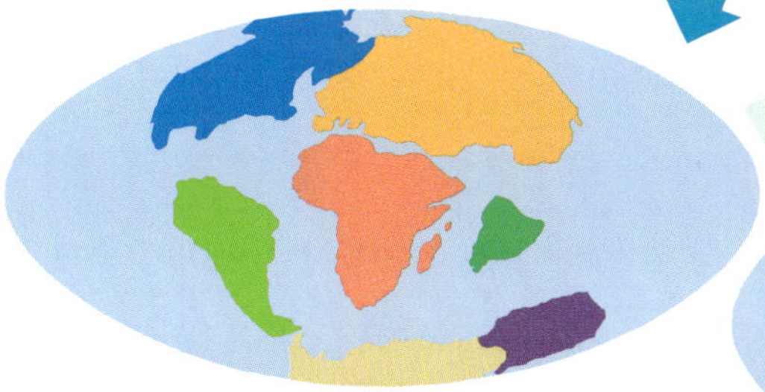

About 65 million years ago, the South Atlantic Ocean was much smaller than it is today.

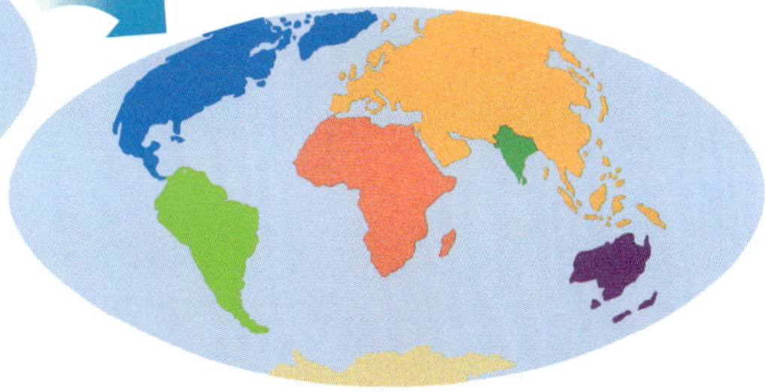

Today the continents continue to move at a rate of 1–10 cm per year. While the Pacific Ocean is getting smaller, the other oceans are expanding.

Scientists at Odds

When Earth cooled about 4 billion years ago, the rains that resulted lasted for thousands of years. But some scientists do not believe all the water on Earth came from condensation as Earth cooled. Instead, they argue that some of the water came from "cosmic rain"—comets that struck Earth in its early history. Encourage students to find out more about this debate.

Answer to Self-Check

If North America and South America continue to drift westward and Asia continues to drift eastward, the continents will eventually collide on the other side of the Earth.

2 Teach

Reading Strategy

Mnemonics Encourage students to create a mnemonic device that will help them remember the names of the world's oceans. For example, they might write, **A**unt **P**atty **A**te **I**nchworms in order to recall **A**tlantic, **P**acific, **A**rctic, and **I**ndian. Have students share their ideas with the class.

Misconception Alert

Students might find the terms *sea* and *ocean* confusing. In some cases the words are interchangeable, but other times the terms mean different things. Parts of the global ocean that are partly surrounded by land are known as seas. The Mediterranean Sea is an example of a sea that is part of the global ocean. Seas that are completely landlocked, such as the Caspian Sea, are not part of the global ocean. Have students identify landlocked seas and the global ocean on a map.

Connect to Physical Science

Explain to students that condensation is a physical change from a gas to a liquid. In the atmosphere, the amount of water vapor that the air can hold is dependent on temperature. As temperature decreases, the air can hold less water vapor, and the water vapor condenses, forming clouds. Ask students to consider other examples of condensation in their lives. (Examples might include condensation on the side of a soda can, dew in the morning, or a foggy mirror in the bathroom.)

2 Teach, continued

MISCONCEPTION ALERT

In **Figure 1,** students may notice that the percentages of some of the elements dissolved in ocean water are particularly low. This does not necessarily mean these elements are not abundant in the ocean. Organisms, such as diatoms and coral, remove dissolved minerals containing some of these elements and use them to make hard body parts. Ask students to note the two elements that are most abundant. (sodium and chloride) Ask them what these elements form when they are combined. (sodium chloride or salt)

BRAIN FOOD

To help students conceptualize the concentration of gold in sea water, point out that 1 billion metric tons of ocean water contains about 4 kg of gold. Have students think of strategies to remove gold from sea water. On land, some mining companies use a bacteria called *Thiobacillus ferrooxidans* to extract gold and copper from mine wastes. Have students find out more about this process and whether it could be used to recover gold from the oceans.

Characteristics of Ocean Water

You know that ocean water is different from the water that flows from the faucet of your kitchen sink. For one thing, ocean water is not safe to drink. But there are other characteristics that make ocean water special.

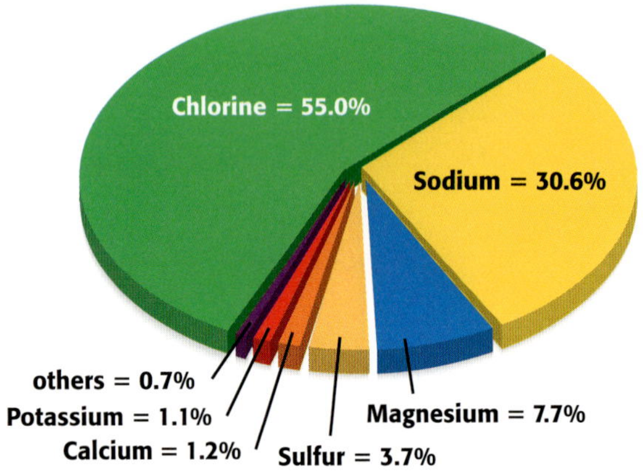

Figure 1 **Percentages of Dissolved Solids in Ocean Water** *This pie chart shows the relative amounts of the dissolved solids in ocean water.*

Ocean Water Is Salty Have you ever swallowed a mouthful of water while swimming in the ocean? It sure had a nasty taste, didn't it? Most of the salt in the ocean is the same kind of salt that we sprinkle on our food. Scientists call this salt *sodium chloride.*

The ocean is so salty because salt has been added to it for billions of years. As rivers and streams flow toward the oceans, they dissolve various minerals on land. The running water carries these dissolved minerals to the ocean. At the same time, water is *evaporating* from the ocean, leaving the dissolved solids behind. The most abundant dissolved solid in the ocean is sodium chloride, a compound of the elements sodium (Na) and chlorine (Cl), as shown in **Figure 1.**

BRAIN FOOD

Did you know that there are about 9 million tons of gold dissolved in the ocean? Too bad the gold's concentration is only 0.000004 mg per kilogram of sea water. Mining the gold from the water would be difficult, and the cost of removing it would be greater than the gold's value.

Chock-full of Solids If more water evaporates than enters the ocean, the ocean's salinity increases. **Salinity** is a measure of the amount of dissolved salts and other solids in a given amount of liquid. Salinity is usually measured as grams of dissolved solids per kilogram of water. Think of it this way: 1 kg (1,000 g) of ocean water contains 35 g of dissolved solids on average. Therefore, if you evaporated 1 kg of ocean water, about 35 g of solids would remain.

Factors That Affect Salinity Some areas of the ocean are saltier than others. Coastal water in areas with hotter, drier climates typically have a higher salinity than coastal water in cooler, more humid areas. This is because less fresh water runs into the ocean in drier areas and because heat increases the evaporation rate. Evaporation removes water but leaves salts and other dissolved solids behind. Also, coastal areas where major rivers run into the ocean have a relatively low salinity. In these areas, the rivers add to the ocean large volumes of fresh water, which contains fewer dissolved solids than sea water.

SCIENCE HUMOR

Q: Why is the ocean salty?

A: because fish don't like pepper

Another factor that affects ocean salinity is water movement. Surface water in some areas of the ocean, such as bays, gulfs, and seas, circulates less than surface water in other parts. Areas in the open ocean that have no currents running through them can also be slow moving. **Figure 2** shows how salinity variations relate to many factors.

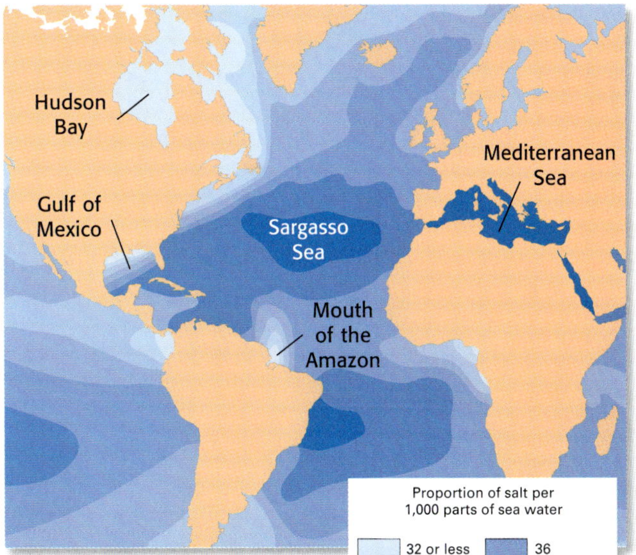

Figure 2 *Salinity varies in different parts of the ocean because of variations in evaporation, circulation, and freshwater inflow.*

Temperature Zones The temperature of ocean water decreases as the depth of the water increases. However, this does not occur gradually from the ocean's surface to its bottom. Water in the ocean can be divided into three layers according to temperature. As you can see in the graph below, the water at the top is much warmer than the average temperature of the ocean.

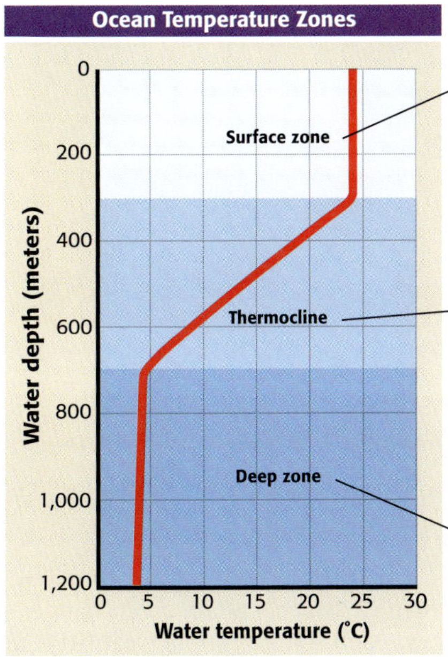

Surface zone
The surface zone is the warm, top layer of ocean water that extends to 300 m below sea level. Sunlight heats the top 100 m of the surface zone. Surface currents mix the heated water with cooler water below.

Thermocline
The **thermocline** is a layer of water extending from 300 m below sea level to about 700 m below sea level. In this zone, water temperature drops with increased depth faster than it does in the other two zones.

Deep zone
This bottom layer extends from the base of the thermocline to the bottom of the ocean. The temperature in this zone averages a chilling 2°C.

MEETING INDIVIDUAL NEEDS

Learners Having Difficulty To help students visualize the temperature zones of the ocean, ask them to make cross-section diagrams of the ocean, labeling the surface zone, thermocline, and deep zone. Have them write captions for their drawings that explain the relationship between temperature and depth. Sheltered English

CONNECT TO PHYSICAL SCIENCE

Sound waves travel faster in warm water than in cold water. A technique called acoustic thermometry of ocean climate (ATOC) measures the time it takes for sound to travel a known distance through the ocean. Using this method, oceanographers can determine the average temperature of the ocean water with great accuracy.

MISCONCEPTION ALERT

Although the graph of ocean temperature zones only extends to a depth of 1,200 m, the ocean is much deeper than 1,200 m in most places, and the deep zone extends all the way to the bottom of the ocean. The rate at which temperature decreases with depth is constant throughout the deep zone. Students should also realize that the depth of each ocean zone varies according to the depth of the ocean floor.

IS THAT A FACT!

The deepest water in the ocean is colder than 0°C, but it remains liquid because of its salinity and the increased pressures at that depth.

WEIRD SCIENCE

The Amazon River feeds so much fresh water into the Atlantic Ocean that the ocean water has different salinity and color almost 160 km from shore!

Teaching Transparency 150
"Ocean Salinity"

2 Teach, *continued*

Group Activity

Making Models Using balloons and permanent markers, students can model how latitude affects ocean surface temperatures. Have them clearly label the poles and equator on their balloons and indicate surface water temperatures between the two. Have them use colored pens to draw bands around their balloons. They should construct a key for their colors, correlating warmer temperatures with the colors closest to the equator. After they color in the shapes of the continents, students can present their models to the class.

Cross-Disciplinary Focus

PORTFOLIO

Geography Encourage students to use atlases or globes to locate the four main oceans. Suggest that students draw a map of the oceans that indicates modern or ancient trade routes and illustrate it with drawings of animals that are unique to each ocean. Students might also make a chart in which they compare all of the oceans by size, average temperature, depth, and other characteristics.

internetconnect

SCILINKS NSTA

TOPIC: Exploring Earth's Oceans
GO TO: www.scilinks.org
***sci*LINKS NUMBER:** HSTE305

Surface Temperature Changes Temperatures in the surface zone vary with latitude and the time of year. Surface temperatures range from 1°C near the poles to about 24°C near the equator. Areas of the ocean along the equator are warmer because they receive more sunlight per year than areas closer to the poles. However, the sun's rays in the Northern Hemisphere are more direct during the summer than during the winter. Therefore, the surface zone absorbs more thermal energy during the summer.

If you live near the coast, you may know firsthand how different a dip in the ocean feels in December than it feels in July. **Figure 3** shows how surface-zone temperatures vary depending on the time of year.

Figure 3 *These satellite images show that the surface temperatures in the northern Pacific Ocean change with the seasons.*

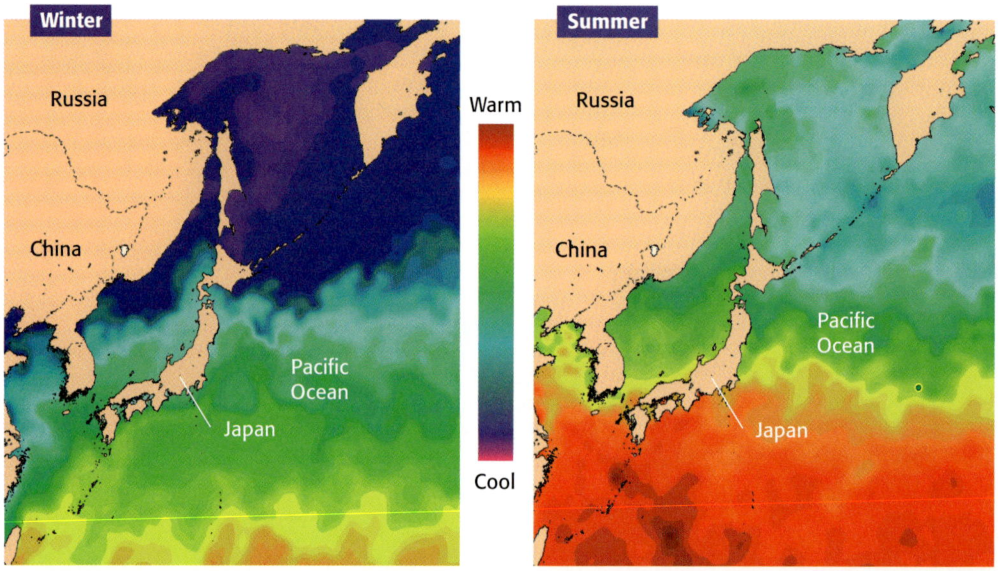

internetconnect

SCILINKS NSTA

TOPIC: Exploring Earth's Oceans
GO TO: www.scilinks.org
***sci*LINKS NUMBER:** HSTE305

SECTION REVIEW

1. Name the major divisions of the global ocean.
2. Explain how Earth's first oceans formed.
3. **Summarizing Data** List three factors that affect salinity in the ocean and three factors that affect ocean temperatures. Explain how each factor affects salinity or temperature.

Answers to Section Review

1. The major divisions of the global ocean are the Pacific Ocean, the Atlantic Ocean, the Indian Ocean, and the Arctic Ocean.
2. Sometime before 4 billion years ago, the Earth cooled enough for water vapor in the atmosphere to condense and fall as rain. The rain began filling the lower levels of Earth's surface, and the first oceans began to form.
3. The factors affecting salinity are the type of climate, the addition or removal of fresh water, and water movement; the factors affecting temperature are water depth, latitude, and the time of year. Explanations will vary.

The Ocean and the Water Cycle

If you could sit on the moon and look down at Earth, what would you see? You would notice that Earth's surface is made up of three basic components—water, land, and air. All three are involved in an ongoing process called the water cycle, as shown below. The **water cycle** is a cycle that links all of Earth's solid, liquid, and gaseous water together. The ocean is an important part of the water cycle because nearly all of Earth's water is found in the ocean.

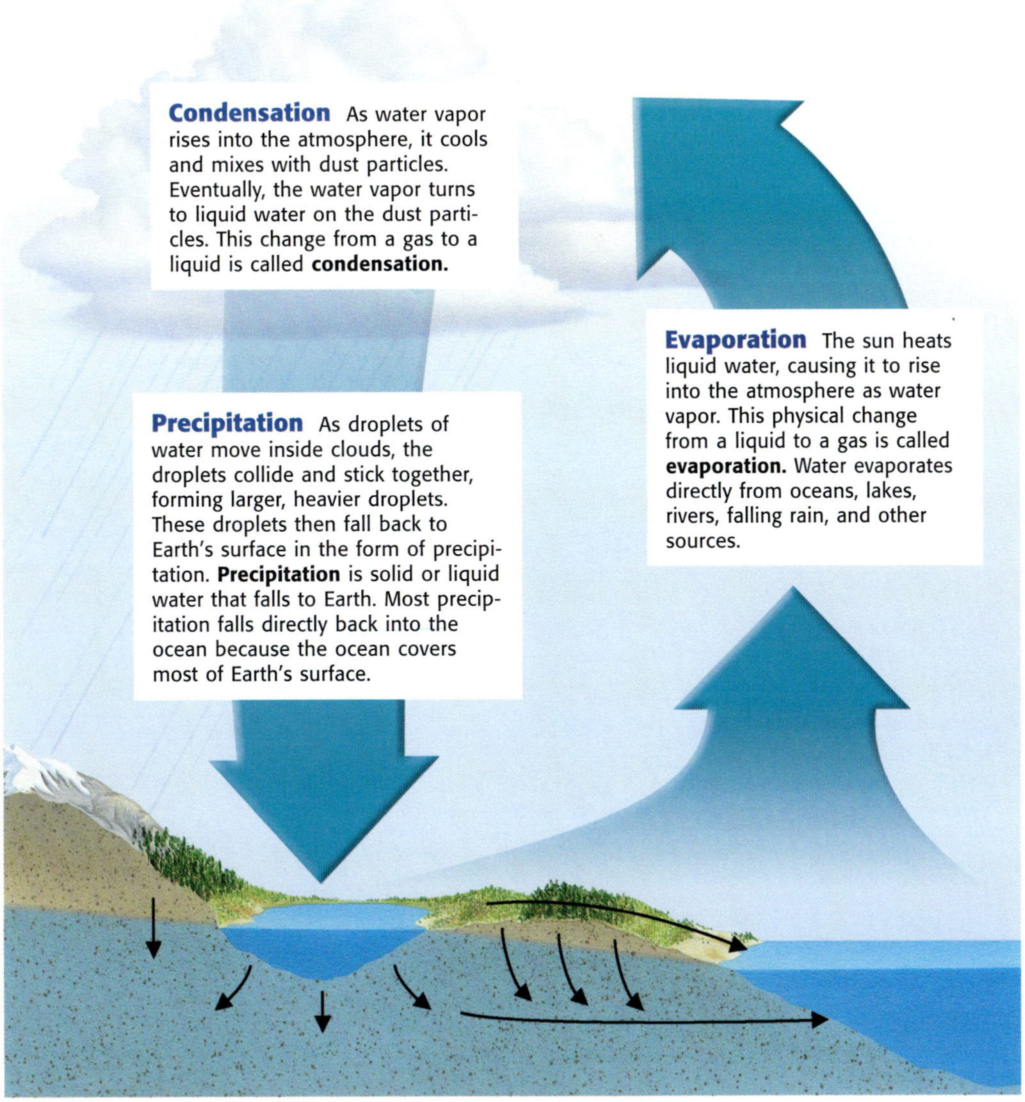

39

IS THAT A FACT!

Twelve thousand years ago much of Earth's water was frozen in glaciers and icecaps, and the Atlantic coast was miles farther out than it is today. Modern-day divers exploring the Chesapeake Bay found a mound of oyster shells—the remains of a long-ago picnic—40 m below sea level!

Teaching Transparency 151
"The Ocean and the Water Cycle"

3 Extend

Meeting Individual Needs

Advanced Learners Divide the class into small groups, and provide each with a small dish or bowl, a plastic bag with twist-tie, and water. Challenge them to create a model demonstrating the water cycle. (Students might place the dish, filled with water, into the bag and seal it. As water evaporates from the dish, it will condense on the inner surface of the bag, eventually falling as "rain." Allow time for each group to explain its model to the class.)

Going Further

Challenge students to explain why the oceans are a crucial influence on the world's weather. (The oceans receive and absorb a large portion of the sun's energy. This energy raises the temperature of the oceans, which influences the atmosphere. Winds create currents, which carry warm water to colder areas, and vice versa, affecting local climates. Finally, solar energy evaporates an enormous amount of sea water, which eventually returns to Earth as precipitation.)

Multicultural Connection

About 3,000 years ago, Greek maps of the world had the Mediterranean Sea in the center of a flat world. Oceans surrounded the lands around the Mediterranean Sea. Ask students: Why do you think the Greeks drew the Mediterranean Sea at the center of their maps? (Their known world centered around it.)

In what ways were their maps considered accurate? (The Greek maps had fairly accurate details of local coastlines and topography.)

4 Close

Quiz

1. How is the global ocean divided, and what are the divisions? (It is divided by the continents into four main oceans, the Pacific, the Atlantic, the Indian, and the Arctic.)
2. How do scientists think the oceans are likely to change in the future? (They predict that the oceans will change in size and shape as the continents move apart.)

Alternative Assessment

Encourage students to examine the illustration on page 35. Have them prepare a timeline detailing the history of Earth's oceans. Challenge students to predict how the oceans will change during the next 150 million years. Students can illustrate their timelines with drawings of each stage of Earth's history.

Misconception Alert

Sea level is not the same worldwide for several reasons. Tides alter the ocean's depth constantly and so do winds. Pacific Ocean trade winds blow westward, causing the ocean level to be about a half meter higher on the western side of the Pacific. Sea level is also higher at the equator than at the poles because the warm equatorial waters expand and the centrifugal force of Earth's rotation causes the middle of the planet to bulge.

A Global Thermostat

The ocean plays a vital role in maintaining conditions favorable for life on Earth. Perhaps the most important function of the ocean is to absorb and hold energy from sunlight. This function regulates temperatures in the atmosphere.

A Hot Exchange The ocean absorbs and releases thermal energy much more slowly than dry land does. If it were not for this function of the ocean, the average air temperature on Earth would vary from above 100°C during the day to below –100°C at night. This rapid exchange of energy between the atmosphere and the Earth's surface would cause violent weather patterns. Life as we know it could not exist with these unstable conditions.

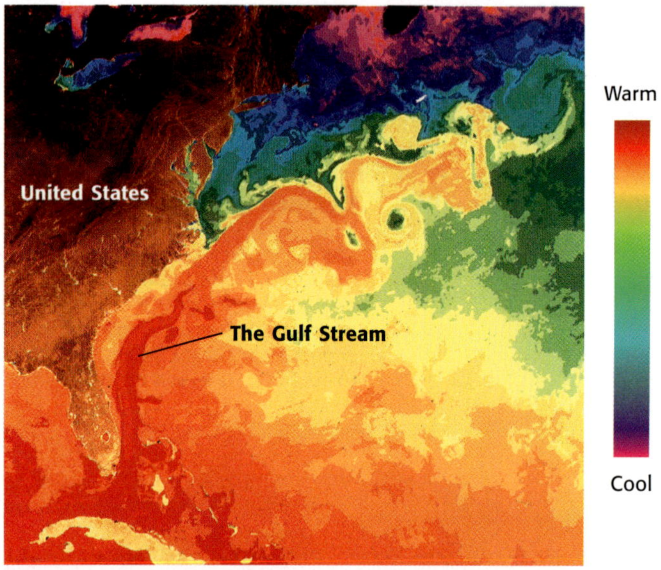

Figure 4 *This infrared satellite image shows the Gulf Stream moving warm water from lower latitudes to higher latitudes.*

Have Heat, Will Travel The ocean also regulates temperatures on a more local scale. At the equator, the sun's rays are more direct, which causes equatorial waters to be warmer than waters at higher latitudes. But currents in the oceans circulate water, as well as the energy it contains, as shown in **Figure 4.** This circulation of warm water causes some coastal lands to have warmer climates than they would have without the currents. The British Isles, for example, have a warmer climate than most regions at the same latitude because of the warm water of the Gulf Stream.

Section Review

1. Why is the ocean an important part of the water cycle?
2. Between which two steps of the water cycle does the ocean fit?
3. **Making Inferences** Explain why St. Louis, Missouri, has colder winters and warmer summers than San Francisco, California, even though the two cities are at about the same latitude.

40

Answers to Section Review

1. The ocean is an important part of the water cycle because nearly all of Earth's water is found in the ocean. In addition, students might note that the ocean absorbs the majority of the solar radiation that reaches the Earth.
2. The ocean fits between precipitation and evaporation in the water cycle.
3. San Francisco is on the coast of the Pacific Ocean. Because the coastal waters absorb and slowly release energy from sunlight, San Francisco has less severe temperature fluctuations. Although St. Louis is at about the same latitude as San Francisco, there is no ocean nearby to regulate temperatures. Thus St. Louis has warmer summers and colder winters than San Francisco.

SECTION 2 READING WARM-UP

The Ocean Floor

Terms to Learn

continental shelf
continental slope
continental rise
abyssal plain
mid-ocean ridge
rift valley
seamount
ocean trench

What You'll Do

- Identify the two major regions of the ocean floor.
- Classify subdivisions and features of the two major regions of the ocean floor.
- Describe technologies for studying the ocean floor.

What lies at the bottom of the ocean? How deep is the ocean? These are questions that were once unanswerable. But humans have learned a lot about the ocean floor, especially in the last few decades. Using state-of-the-art technology, scientists have discovered a wide variety of landforms on the ocean floor. Scientists have also determined accurate depths for almost the entire ocean floor.

Exploring the Ocean Floor

Some parts of the ocean are so deep that humans must use special underwater vessels to travel there. Perhaps the most familiar underwater vessel used by scientists to study the ocean floor is the minisub called *Alvin.* Scientists have used *Alvin* for many underwater missions, including searches for sunken ships, the recovery of a lost hydrogen bomb, and explorations of landforms on the sea floor.

Although the use of *Alvin* has enabled scientists to make some amazing discoveries, scientists are developing new vessels for ocean exploration, such as an underwater airplane called *Deep Flight.* This vessel, shown in **Figure 5,** moves through the water much like an airplane moves through the air. Future models of *Deep Flight* will be designed to transport pilots to the deepest part of the ocean.

Science CONNECTION

Turn to page 66 to meet the most famous underwater explorer who ever lived.

Figure 5 *Like the Wright brothers' first successful airplane,* Deep Flight *sets the stage for a bright future—this time in underwater "flight."*

41

IS THAT A FACT!

The same explorer who led the first voyage around the world also attempted to determine the depth of the ocean. In 1520, Ferdinand Magellan weighted a 370 m rope with lead, and lowered it into the ocean. Unfortunately, his rope was not long enough to reach the ocean floor. It was not until 1773 that a successful measurement was made. Using Magellan's techniques, explorers found that the depth of the ocean near Norway is about 1,250 m.

SECTION 2

Focus

The Ocean Floor

In this section, students learn about the regional divisions of the ocean floor and the geographic features of each division. They learn how technology has facilitated exploration of the ocean floor, and they learn about the methods used to survey the ocean floor, including sonar and satellites.

Bellringer

Before students read this section, have them explore their ideas about the ocean floor by telling them to pretend they have walked off the edge of North America and into the depths of the Atlantic Ocean. As they walk along the ocean floor toward Europe, what would they see? Have each student make a drawing of the ocean floor that he or she would see along the way. Sheltered English

1 Motivate

DISCUSSION

It has been said that scientists know more about the surface of the moon than about the ocean floor. Most of what scientists know about the ocean floor comes from sonar readings and sample dredging. Ask students to think about why the ocean is so difficult to study and what kinds of technology would help scientists learn more about the deep-ocean floor.

Directed Reading Worksheet Section 2

2 Teach

MATH and MORE

Water pressure increases with depth. For every 10 m of depth, the pressure increases by 1 atmosphere (atm). For example, at a depth of 10 m, the pressure is 2 atm, or twice the pressure of the atmosphere at sea level. Ask students:

What is the pressure at 20 m? (3 atm)

50 m? (6 atm)

100 m? (11 atm)

What is the pressure at the bottom of the Mariana Trench? (greater than 1200 atm)

Math Skills Worksheet "Multiplying Whole Numbers"

CONNECT TO PHYSICAL SCIENCE

To show how water pressure changes with depth, take a milk carton and punch three holes in its side: one near the top, one halfway down the side, and one near the bottom. Put one piece of tape over all three holes, and fill the carton with water. Remove the tape quickly. Have students observe the streams of water and explain what they saw. (The water stream at the bottom of the carton shot out the farthest, with the greatest force. This is because the water above exerted pressure on the water at the bottom of the carton.)

Revealing the Ocean Floor

What if you were an explorer assigned to map uncharted areas on the planet? You might think there were not many uncharted areas left because most of the land had already been explored. But what about the bottom of the ocean? If you could travel to the bottom of the ocean in *Deep Flight,* you would see the world's largest mountain chain and canyons deeper than the Grand Canyon. And because it is under water, much of this area is unexplored.

As you began your descent into the underwater realm, you would notice two major regions—the *continental margin,* which is made of continental crust, and the *deep-ocean basin,* which is made of oceanic crust. It may help to imagine the ocean as a giant swimming pool; the continental margin is the shallow end and slope of the pool, and the deep-ocean basin is the deep end of the pool. **Figure 6** shows how these two regions are subdivided.

Figure 6 *The continental margin is subdivided into three depth zones, and the deep-ocean basin consists of one depth zone with several features.*

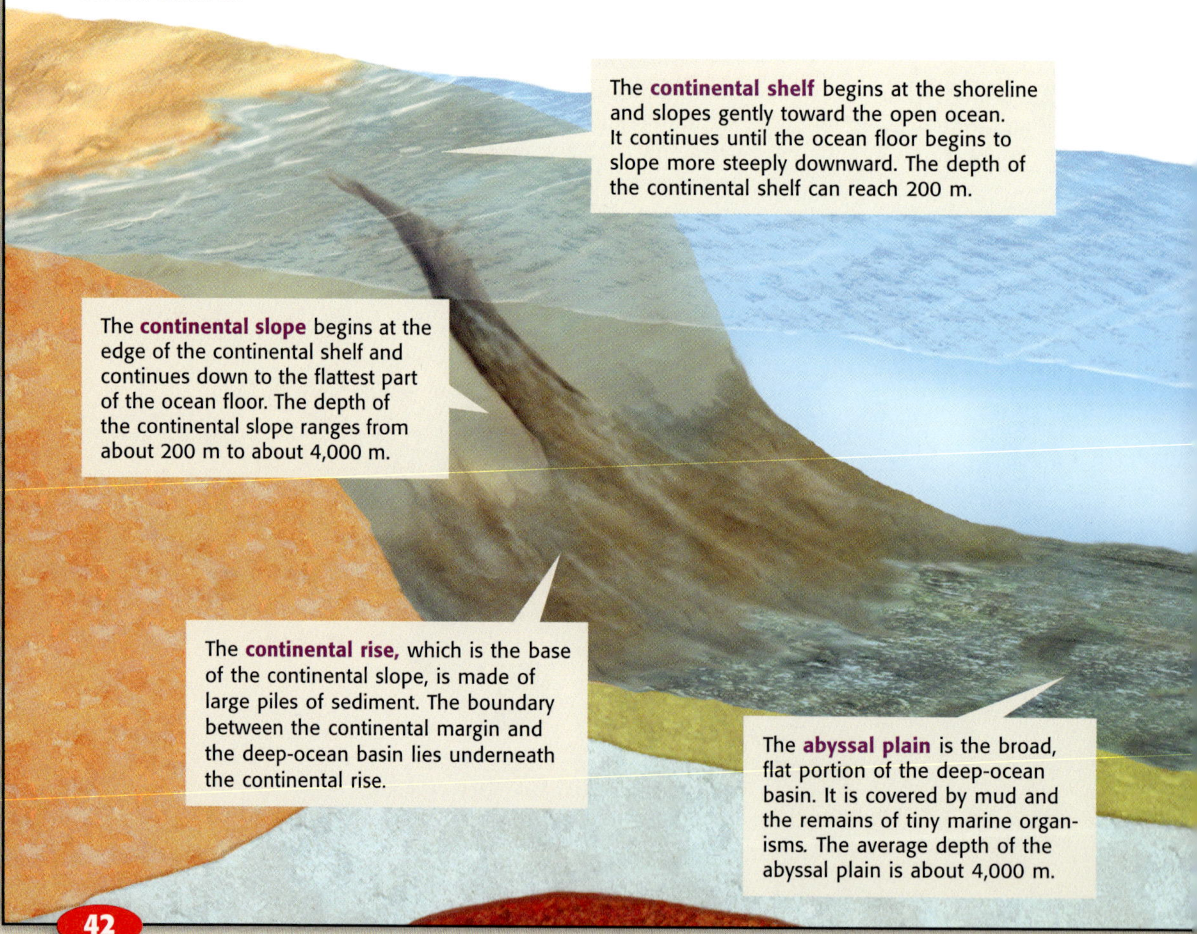

Teaching Transparency 152 "Revealing the Ocean Floor: A"

Teaching Transparency 153 "Revealing the Ocean Floor: B"

SCIENCE HUMOR

Q: What lies on the bottom of the ocean and trembles?

A: a nervous wreck

Underwater Real Estate As you can see, the continental margin is subdivided into the continental shelf, the continental slope, and the continental rise based on depth and changes in slope. The deep-ocean basin consists of the abyssal plain, with features such as mid-ocean ridges, rift valleys, and ocean trenches that form near the boundaries of Earth's *tectonic plates.* On parts of the abyssal plain that are not near plate boundaries, thousands of seamounts are found on the ocean floor.

Self-Check

How do the locations of rift valleys and ocean trenches differ? *(See page 136 to check your answer.)*

Activity

To get an idea of how deep parts of the ocean are, use an encyclopedia to find out how deep the Grand Canyon is. Compare this depth with that of the Mariana Trench, which is more than 11,000 m deep! Make a model of this difference using clay, or draw a graph of this difference to scale.

TRY at HOME

internetconnect

SCILINKS NSTA

TOPIC: The Ocean Floor
GO TO: www.scilinks.com
***sci*LINKS NUMBER:** HSTE310

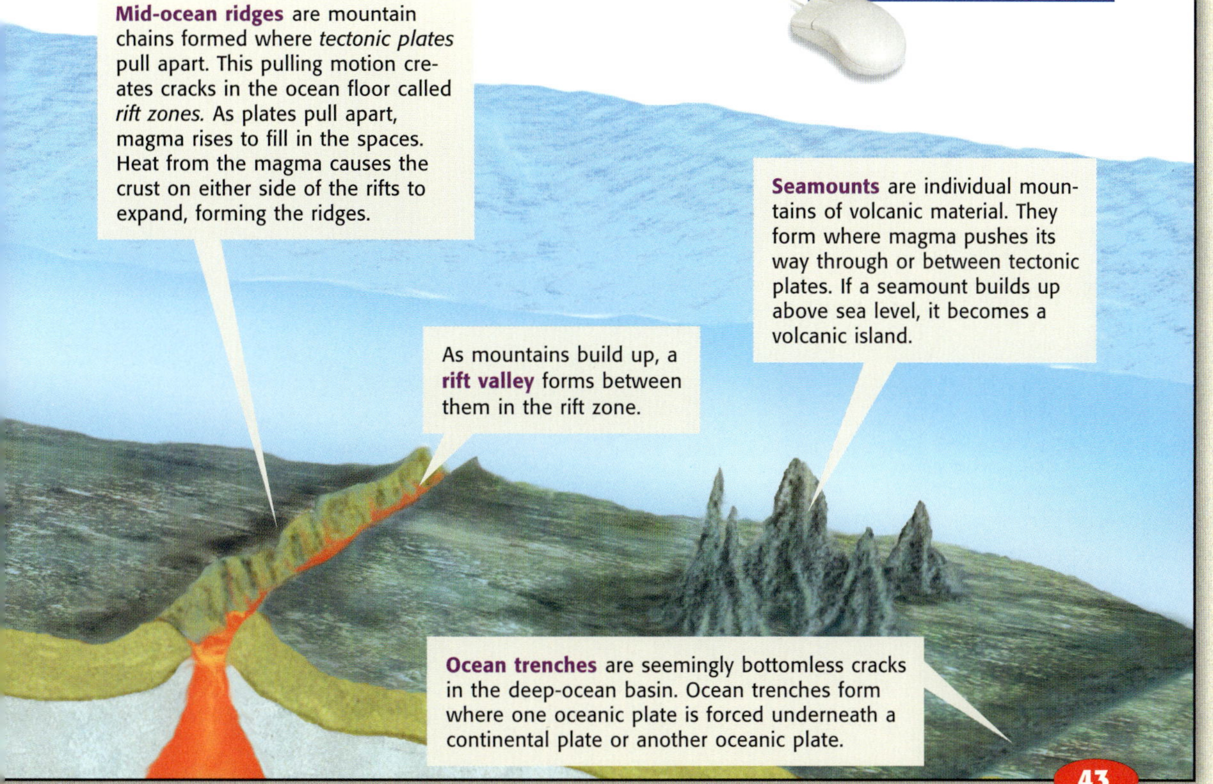

Mid-ocean ridges are mountain chains formed where *tectonic plates* pull apart. This pulling motion creates cracks in the ocean floor called *rift zones.* As plates pull apart, magma rises to fill in the spaces. Heat from the magma causes the crust on either side of the rifts to expand, forming the ridges.

As mountains build up, a **rift valley** forms between them in the rift zone.

Seamounts are individual mountains of volcanic material. They form where magma pushes its way through or between tectonic plates. If a seamount builds up above sea level, it becomes a volcanic island.

Ocean trenches are seemingly bottomless cracks in the deep-ocean basin. Ocean trenches form where one oceanic plate is forced underneath a continental plate or another oceanic plate.

43

Cross-Disciplinary Focus

PORTFOLIO

Art Draw students' attention to **Figure 6,** and have them draw, label, and color their own picture of the depth zones of the ocean floor. Point out the canyon in the continental slope. This is a *submarine canyon.* Most of the sediment that makes up the continental rise travels down from the continental shelf through submarine canyons. Be sure that students have divided the continental margin into the continental shelf, continental slope, and continental rise. They should identify the features of the deep-ocean basin as mid-ocean ridges, seamounts, rift valleys, and ocean trenches. In addition, students can indicate the temperature of the ocean water at each depth by using different colors. Sheltered English

Using the Figure

Draw students' attention to the mid-ocean ridges in **Figure 6.** Below the rift zones that characterize mid-ocean ridges, magma rises from below the crust. As the magma erupts as lava, it cools when it enters the water, forming new oceanic crust. Point out that ocean trenches formed by the subduction of plates are some of the deepest places on Earth and often support a diversity of benthic life.

Homework

PORTFOLIO

Concept Mapping Remind students that volcanic seamounts that rise above the ocean surface become volcanic islands. The Hawaiian Islands formed this way. Have them research other ways islands form. Have students prepare a concept map of the different ways that islands form. They should find that some islands are formed by the growth of coral; some are formed by deposition (barrier islands); and some are continental islands (Great Britain, Madagascar).

Answer to Self-Check

Rift valleys form where tectonic plates pull apart, and ocean trenches form where one oceanic plate is forced underneath a continental plate or another oceanic plate.

internetconnect

SCILINKS NSTA

TOPIC: The Ocean Floor
GO TO: www.scilinks.org
***sci*LINKS NUMBER:** HSTE310

3 Extend

RETEACHING

Have students review **Figure 6,** which shows zones and features of the ocean floor. Ask students to choose any two zones or features and describe them in their own words. If describing features, students should include how the features form. Students should also tell why they like each feature or zone. **Sheltered English**

CROSS-DISCIPLINARY FOCUS

Language Arts Tell students that the 1870 publication of *Twenty Thousand Leagues Under the Sea,* by Jules Verne, revived an interest in undersea exploration. Have students read sections of Verne's book aloud in class. Point out that the story inspired engineers to solve the problems plaguing submarines, enabling scientists to reach greater depths in their exploration of the sea.

Answers to MATHBREAK

1. $D = 1{,}500$ m
2. $D = 10{,}500$ m
3. $D = 3{,}975$ m

Explain why the constant $\frac{1}{2}$ is in the equation in the MATHBREAK. The time represented by t is the two-way travel time of sound waves. Without using the constant $\frac{1}{2}$, the equation would give the distance to the ocean floor and back to the ship instead of the distance to the floor only.

Math Skills Worksheet "Multiplying and Dividing Fractions"

MATH BREAK

Depths of the Deep

The depths in a bathymetric profile are calculated using the following simple formula:

$$D = \frac{1}{2}t \times v$$

D is the depth of the ocean floor, t is the time it takes for the sound to reach the bottom and return to the surface, and v equals the speed of sound in water (1,500 m/s). Calculate D for the following three parts of the ocean floor:

1. a mid-ocean ridge ($t = 2$ s)
2. an ocean trench ($t = 14$ s)
3. an abyssal plain ($t = 5.3$ s)

Viewing the Ocean Floor from Above

In spite of the great success of underwater exploration, sending scientists into deep water is still risky. Fortunately, there are ways to survey the underwater realm from the surface and from high above in space. Read on to learn about two technologies—sonar and satellites—that enable scientists to study the ocean floor without going below the surface.

Seeing by Sonar *Sonar,* which stands for "sound navigation and ranging," is a technology based on the echo-ranging behavior of bats. Scientists use sonar to determine the ocean's depth by sending high-frequency sound pulses from a ship down into the ocean. The sound travels through the water, bounces off the ocean floor, and returns to the ship. The deeper the water is, the longer the round trip takes. Scientists then calculate the depth by multiplying half the travel time by the speed of sound in water (about 1,500 m/s). This process is shown in the illustration below.

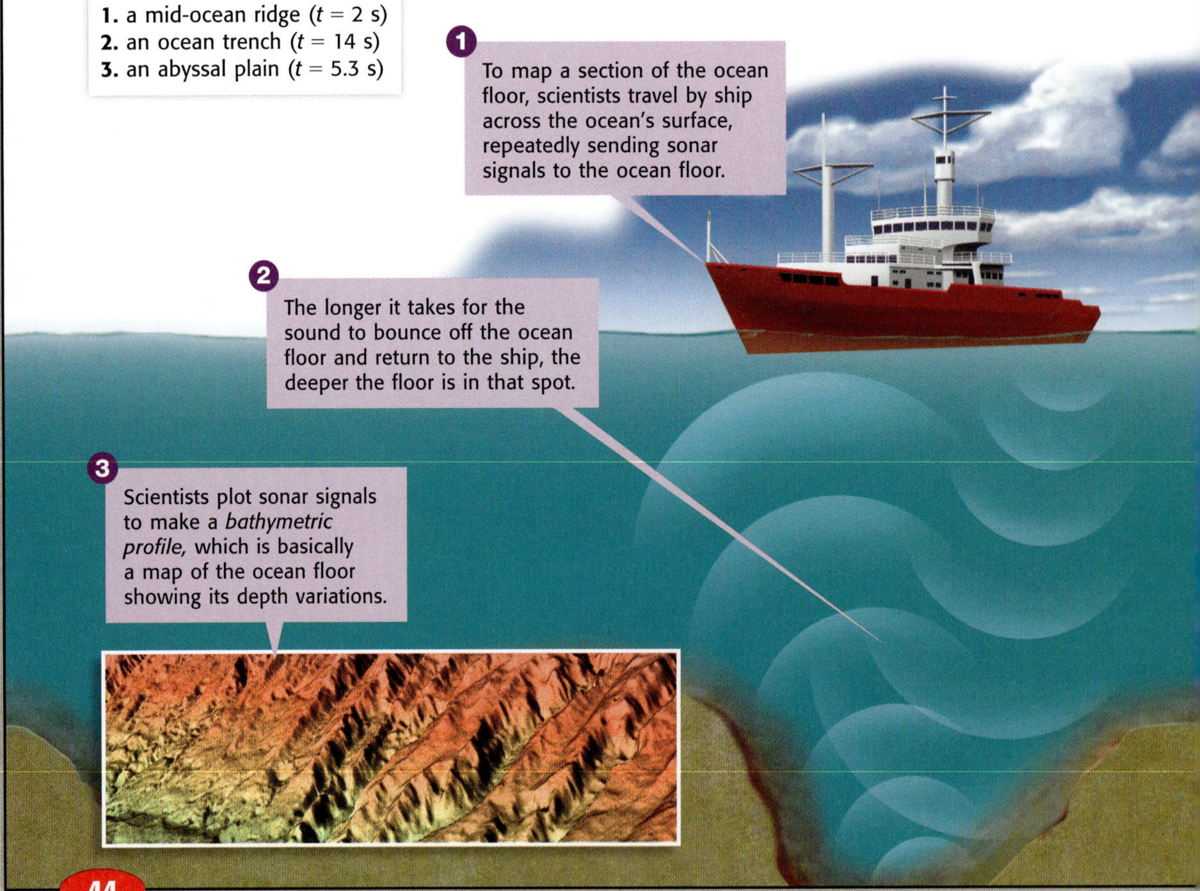

44

Teaching Transparency 154 "How Sonar Works"

WEIRD SCIENCE

Though many people know that whales and dolphins communicate by sound, few are aware that shrimp do the same thing. In order to locate food sources and other shrimp, they emit a sound similar to that of bacon frying!

Oceanography via Satellite In the 1970s, scientists began studying Earth from satellites in orbit around the Earth. In 1978, scientists launched the satellite *Seasat*. This satellite focused on the ocean, sending images back to Earth that allowed scientists to measure the direction and speed of ocean currents.

Geosat, once a top-secret military satellite, has been used to measure slight changes in the height of the ocean's surface. Different underwater features, such as mountains and trenches, affect the height of the water above them, thus reflecting the underwater topography of the ocean floor. Scientists measure the different heights of the ocean surface and use the measurements to make highly detailed maps of the ocean floor. As illustrated in **Figure 7,** oceanographers can make maps that cover a lot more territory by using satellites than by using ship-based sonar readings.

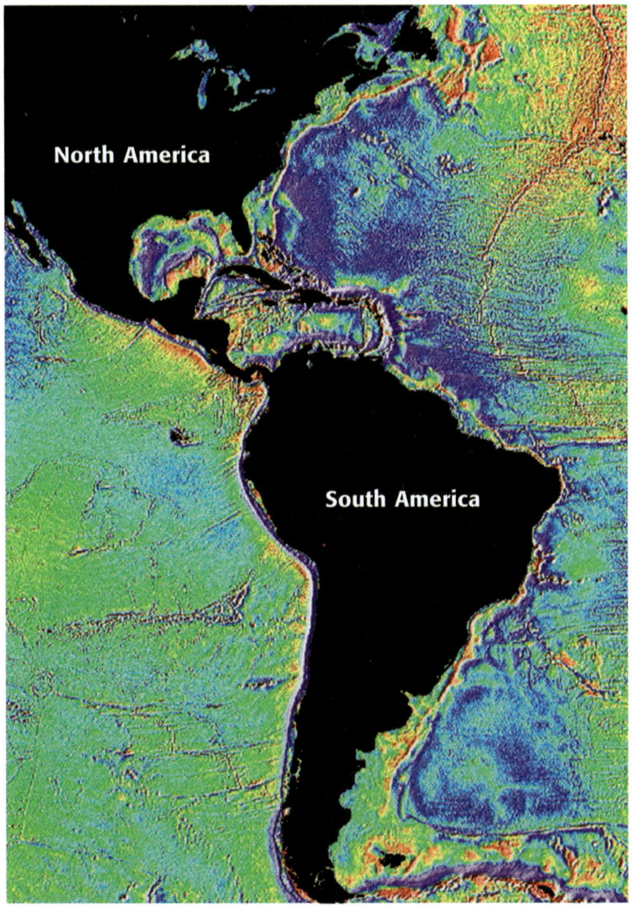

Figure 7 *The map above was generated by satellite measurements of different heights of the ocean surface.*

SECTION REVIEW

1. Name the two major regions of the ocean floor.
2. List the subdivisions of the continental margin.
3. List three technologies for studying the ocean floor, and explain how they are used.
4. **Interpreting Graphics** What part of the ocean floor would the bathymetric profile at right represent?

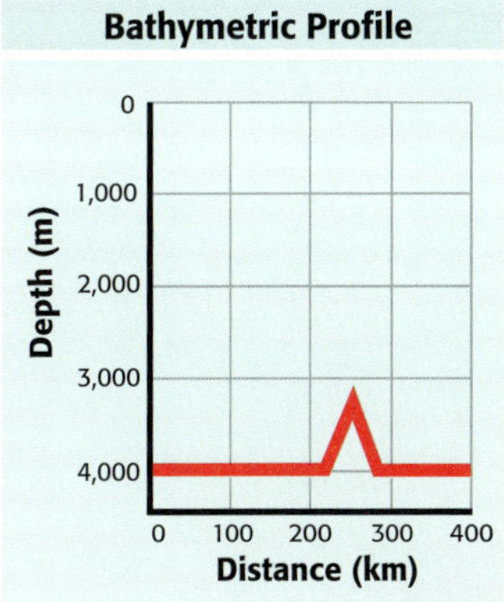

Answers to Section Review

1. the continental margin and the deep-ocean basin
2. the continental shelf, continental slope, and continental rise
3. Answers may include underwater vessels, such as *Alvin* and *Deep Flight,* sonar, and satellites. (Some technologies not mentioned in the text are acceptable.) Scientists visit the ocean floor in underwater vessels. Sonar is used to determine the ocean's depth and the topography of the ocean floor. Satellites are used to measure the direction and speed of ocean currents and measure changes in the height of the ocean surface that reflect the topography of the ocean floor.
4. The abyssal plain; the sharp rise and fall in the profile represents a seamount or part of a mid-ocean ridge.

BRAIN FOOD

Students may be surprised to learn that underwater features such as seamounts and trenches affect the height of the ocean above them. Interestingly, the topography and composition of the ocean floor causes differences in gravity that are pronounced enough to affect sea level. For example, the mass of a seamount creates enough gravitational attraction to cause a 5 m bulge of water above it. Similarly, the sea level above a deep-sea trench can be depressed by as much as 60 m!

4 Close

Quiz

1. Which features of the abyssal plain form at the boundaries of tectonic plates? (mid-ocean ridges, rift valleys, ocean trenches)
2. How is the depth of the ocean measured? (It is measured using sonar. Scientists calculate the depth by multiplying half the time it takes a sound wave to hit the ocean floor and return to the surface by the speed of sound in water.)

ALTERNATIVE ASSESSMENT

Encourage students to imagine that they are oceanographers on a deep-sea mission aboard *Alvin.* Have them work in groups to make a model of the ocean floor, including all the features from this section. The model can be a cross section similar to **Figure 6** or a view from above.

Sheltered English

SECTION 3

Focus

Life in the Ocean

This section introduces a system for classifying marine organisms based on where they live and how they move. Students also learn to describe ecological zones of the ocean and give examples of organisms inhabiting each zone.

Bellringer

Before they read this section, have students imagine they are marine biologists who must classify marine life into three groups. Challenge them to identify the criteria they would use in their classification systems.

1 Motivate

GROUP ACTIVITY

Classifying Divide the class into small groups. Ask them to classify as many items in the classroom as they can based on the following categories:

- height at which the items are located
- what the items are used for

Teaching Transparency 155 "The Three Groups of Marine Life"

Directed Reading Worksheet Section 3

SECTION 3 READING WARM-UP

Terms to Learn

plankton
nekton
benthos
benthic environment
pelagic environment

What You'll Do

- Identify and describe the three groups of marine organisms.
- Identify and describe the benthic and pelagic environments.
- Classify the zones of the benthic and pelagic environments.

Life in the Ocean

The ocean contains a wide variety of life-forms, many of which we know little about. Trying to study them can be quite a challenge for scientists. To make things easier, scientists classify marine organisms into three main groups. Scientists also divide the ocean into two main environments based on the types of organisms that live in them. These two main environments are further subdivided into ecological zones based on locations of different organisms.

The Three Groups of Marine Life

The three main groups of marine life are plankton, nekton, and benthos. Marine organisms are placed into one of these three groups according to where they live and how they move. Carefully examine the figure below to understand the differences between these groups.

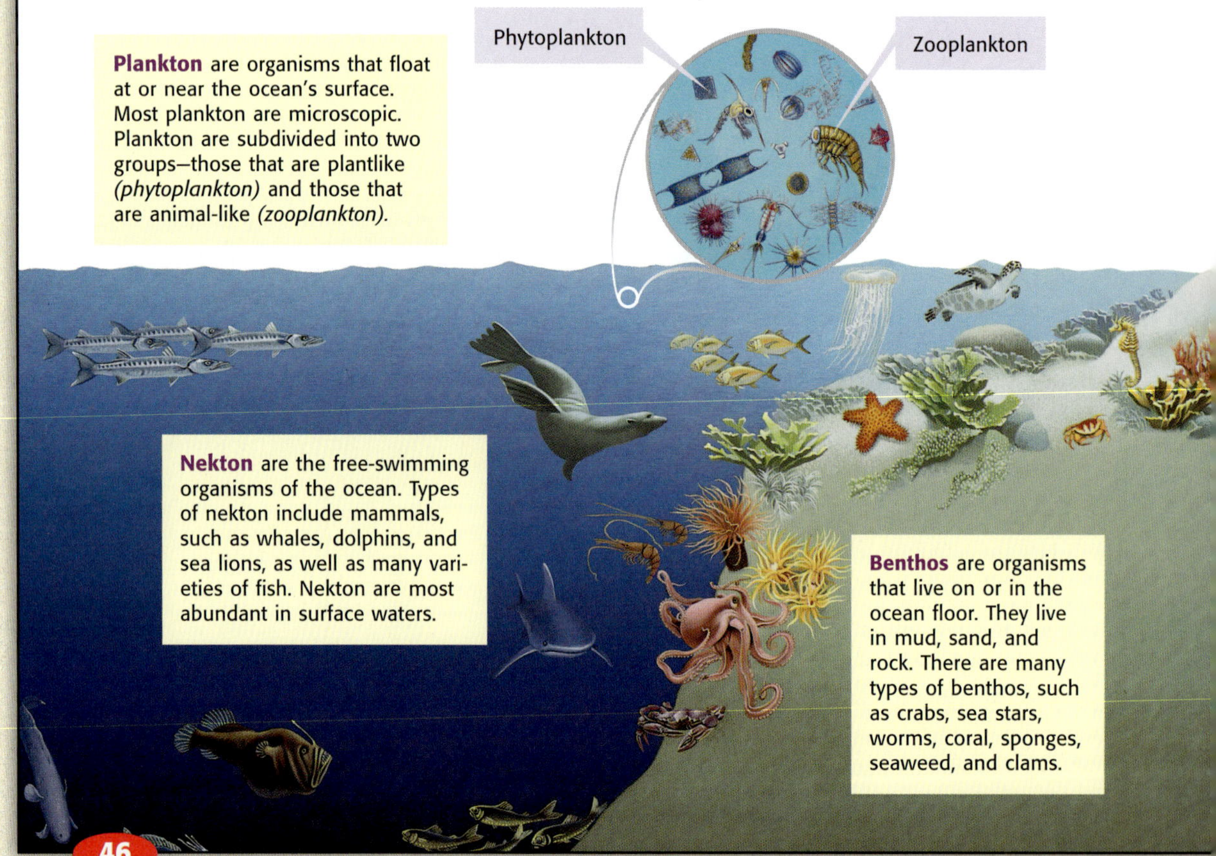

Plankton are organisms that float at or near the ocean's surface. Most plankton are microscopic. Plankton are subdivided into two groups—those that are plantlike *(phytoplankton)* and those that are animal-like *(zooplankton)*.

Nekton are the free-swimming organisms of the ocean. Types of nekton include mammals, such as whales, dolphins, and sea lions, as well as many varieties of fish. Nekton are most abundant in surface waters.

Benthos are organisms that live on or in the ocean floor. They live in mud, sand, and rock. There are many types of benthos, such as crabs, sea stars, worms, coral, sponges, seaweed, and clams.

46

SCIENTISTS AT ODDS

Before the late 1870s, scientists widely believed the "azoic theory" of James Forbes and Alexander Agassiz. This theory argued that no life existed below shallow depths in the oceans. Sir Wyville Thomson disputed this theory, and in 1872, he embarked on a three-and-a-half year voyage to prove his point. The HMS *Challenger* voyage, which established oceanography as a modern science, collected ocean-floor samples from deeper than 8,000 m. Thomson's evidence disproved the azoic theory and greatly expanded our knowledge of the world's oceans. He published his results in *The Depths of the Sea,* the first general textbook on oceanography.

The Benthic Environment

In addition to being divided into zones based on depth, the ocean floor is divided into ecological zones based on where different types of benthos live. These zones are grouped into one major marine environment—the benthic environment. The **benthic environment,** or bottom environment, is the ocean floor and all the organisms that live on or in it.

Intertidal Zone The shallowest benthic zone, the *intertidal zone,* is located between the low-tide and high-tide limits. Twice a day, the intertidal zone transforms. As the tide flows in, the zone is covered with ocean water, and as the tide retreats, the intertidal zone is exposed to the air and sun.

Intertidal organisms must be able to live both underwater and on exposed land. Some organisms attach themselves to rocks and reefs to avoid being washed out to sea during low tide, as shown in **Figure 8.** Clams, oysters, barnacles, and crabs have tough shells that give them protection against strong waves during high tide and against harsh sunlight during low tide. Some animals can burrow in sand or between rocks. Plants such as seaweed have strong *holdfasts* (rootlike structures) that allow them to grow in this zone.

Sublittoral Zone The *sublittoral zone* begins where the intertidal zones ends, at the low-tide limit, and extends to the edge of the continental shelf. This benthic zone is more stable than the intertidal zone; the temperature, water pressure, and amount of sunlight remain fairly constant. Consequently, sublittoral organisms, such as corals, shown in **Figure 9,** do not have to cope with as much change as intertidal organisms. Although the sublittoral zone extends down 200 m below sea level, plants and most animals stay in the upper 100 m, where sunlight reaches the ocean floor.

Biology CONNECTION

Coral reefs, found in shallow marine waters, have the largest concentration of life in the ocean. Layers of skeletons from animals called *corals* form the reefs, which are the largest animal structures on Earth. Many other organisms live on, around, and even in coral reefs.

Figure 8 *Organisms such as sea anemones and starfish attach themselves to rocks and reefs. These organisms must be able to survive wet and dry conditions.*

Figure 9 *Corals, like many other types of organisms, can live in both the sublittoral zone and the intertidal zone. However, they are more common in the sublittoral zone.*

47

Multicultural CONNECTION

Writing

Stories of mermaids and sea monsters abound in many different cultures. In some legends, mermaids were good and helped shipwrecked sailors. In others, mermaids lured ships and sailors into dangerous waters. Sea monsters, on the other hand, were always bad, sinking ships and killing sailors. Discuss with students why they think these legends were told. Have them research a legend and write one of their own.

2 Teach

READING STRATEGY

Mnemonics Ask students to create a mnemonic device that will remind them of the zones of the benthic environment. For example, they might write **I**sabel **S**ent **B**art **A**way from **H**ome to remind them that the zones are **I**ntertidal, **S**ublittoral, **B**athyal, **A**byssal, and **H**adal. Ask them to share their mnemonic devices with the class. Sheltered English

MISCONCEPTION ALERT

Many people think that rain forests produce most of Earth's oxygen. Actually, ocean phytoplankton are the most productive photosynthesizers on the planet. Because phytoplankton consume the greenhouse gas CO_2 during photosynthesis, many scientists think that phytoplankton populations play a significant role in global climate patterns.

Homework

PORTFOLIO

Have students draw five imaginary organisms, each inhabiting a different benthic zone. Students should describe each plant or animal and explain its particular adaptations for living in that zone. Their drawings and descriptions should include the following:

- how they obtain food
- how they avoid predation
- how they withstand the water pressure and temperature at the depth where they live

Encourage students to share their drawings and explanations with the class. Sheltered English

3 Extend

Cooperative Learning

Divide the class into groups of five, and provide each group with poster board and markers. Ask them to draw a cross-sectional illustration of the ocean and label the following features:

benthic environment, intertidal zone, sublittoral zone, bathyal zone, abyssal zone, hadal zone

Have each group member illustrate the organisms found in one of the five zones. (Each group should illustrate all of the five zones.) Instruct each group to elect a spokesperson to present the poster to the class.

Research

PORTFOLIO Have students select one of the benthic zones to investigate further. Ask them to focus on the adaptations that organisms living there have developed that enable them to exist in that zone. Students can present their findings in a concept map, poster, or comic book. Sheltered English

Connect to Life Science

Because food is so scarce in the deeper parts of the ocean, its inhabitants have special adaptations to ensure their survival. Some gulper eels, for example, have huge jaws and elastic stomachs that allow them to eat fish larger than themselves. Show students photographs of some of these organisms, and use Teaching Transparency 27 to discuss the process of natural selection.

Teaching Transparency 27
"Natural Selection in Four Steps"
LINK TO LIFE SCIENCE

Figure 10 *Octopuses are one of the animals common to the bathyal zone.*

Figure 11 *Tube worms can tolerate higher temperatures than most other organisms. These animals survive in water as hot as 81°C.*

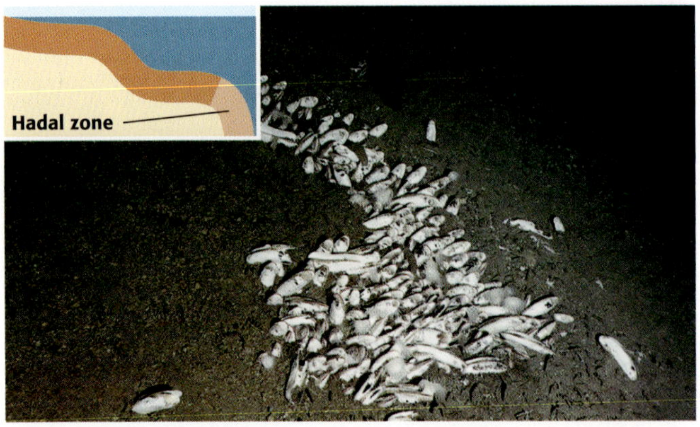

Figure 12 *These clams are one of the few types of organisms known to live in the hadal zone.*

Bathyal Zone The *bathyal zone* extends from the edge of the continental shelf to the abyssal plain. The depth of this zone ranges from 200 m to 4,000 m below sea level. Because of the lack of sunlight at these depths, plant life is scarce in this part of the benthic environment. Animals in this zone include sponges, *brachiopods,* sea stars, *echinoids,* and octopuses, such as the one shown in **Figure 10.**

Abyssal Zone No plants and very few animals live in the *abyssal zone,* which is on the abyssal plain. Among the abyssal animal types are crabs, sponges, worms, and sea cucumbers. Many of these organisms, such as the tube worms shown in **Figure 11,** live around hot-water vents called *black smokers*. The abyssal zone can reach 6,000 m in depth. Scientists know very little about this benthic zone because it is so deep and dark.

Hadal Zone The deepest benthic zone is the *hadal zone.* This zone consists of the floor of the ocean trenches and any organisms found there. Scientists know even less about the hadal zone than they do about the abyssal zone. So far, scientists have discovered a type of sponge, a few species of worms, and a type of clam, which is shown in **Figure 12.**

48

internet connect

SCILINKS NSTA
TOPIC: Life in the Oceans
GO TO: www.scilinks.org
***sci*LINKS NUMBER:** HSTE315

SCIENCE HUMOR

Q: What's the best way to catch a fish?

A: Have someone throw it to you.

The Pelagic Environment

The **pelagic environment** is the entire volume of water in the ocean and the marine organisms that live above the ocean floor. There are two major zones in the pelagic environment—the *neritic zone* and the *oceanic zone.*

Neritic Zone The neritic zone includes the volume of water that covers the continental shelf. This warm, shallow zone contains the largest concentration of marine life. This is due to an abundance of sunlight and to the many benthos below the neritic zone that serve as a food supply. Fish, plankton, and marine mammals, such as the one in **Figure 13,** are just a few of the animal groups found here.

Figure 13 *Many marine mammals, such as this dolphin, live in the neritic zone.*

Oceanic Zone The oceanic zone includes the volume of water that covers the entire sea floor except for the continental shelf. In the deeper parts of the oceanic zone, the water temperature is colder and the pressure is much greater than in the neritic zone. Also, organisms are more spread out in the oceanic zone than in the neritic zone. While many of the same organisms that live in the neritic zone are found throughout the upper regions, some strange animals lurk in the darker depths, as shown in **Figure 14.** Other animals in the deeper parts of this zone include giant squids and some whale species.

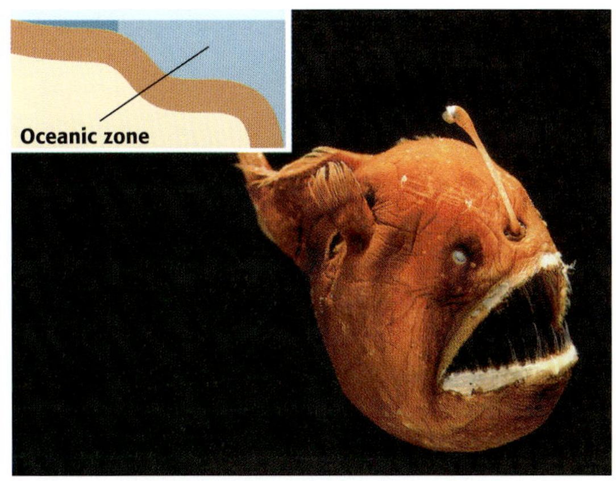

Figure 14 *The anglerfish is a tricky predator that uses a natural lure attached to its head to attract prey.*

SECTION REVIEW

1. List and briefly describe the three main groups of marine organisms.
2. Name the two ocean environments. List the zones of each environment.
3. **Making Predictions** How would the ocean's ecological zones change if sea level dropped 300 m?

internetconnect

SCILINKS. NSTA

TOPIC: Life in Oceans
GO TO: www.scilinks.org
***sci*LINKS NUMBER:** HSTE315

49

CONNECT TO LIFE SCIENCE

Sea skaters are the only insects known to live directly on the surface of the open ocean. The insects have waxy hairs on their feet that repel water and help them travel on the surface of the water. They are predators that hunt prey swimming beneath the surface by spearing them with sharp mouth parts. Ask students to list reasons why other insects do not live in the open ocean.

4 Close

Quiz

1. Why do scientists know little about the abyssal and hadal zones? (Because these zones are so deep and dark, scientists know very little about them.)
2. Where do most marine organisms live? (in the neritic zone)

Alternative Assessment

PORTFOLIO Challenge students to write science-fiction stories about exploring the abyssal zone of the benthic environment. Have them describe the difficulties of exploring that zone, and encourage them to use their imagination to describe creatures they might find there.

Answers to Section Review

1. plankton, nekton, and benthos; Plankton float at or near the surface, nekton swim freely, and benthos live on or in the ocean floor.
2. the benthic environment and the pelagic environment; the zones of the benthic environment are the intertidal, sublittoral zone, bathyal, abyssal, and hadal zones; the zones of the pelagic environment are the neritic and oceanic zones.
3. Answers will vary. If sea level dropped 300 m, the shoreline would move down to the continental slope. This would cause an overall shift of ecological zones as many marine organisms are restricted to certain depths.

Reinforcement Worksheet "The Ocean's Environment"

Interactive Explorations CD-ROM "Sea Sick"

SECTION 4

Focus

Resources from the Ocean

This section discusses the ocean's living and nonliving resources, focusing on the methods used to obtain them. Students are asked to consider the importance of the ocean's resources and explore ways of managing and conserving them.

Bellringer

Before they read this section, write the following sentences on the board and challenge students to identify four items or activities that involve resources that can be obtained from the sea.

"Tabitha drove her car to the market to buy a tuna steak for dinner. When she got home, she poured herself a glass of water, then fired up her gas grill, and cooked the tuna."

1) Motivate

GROUP ACTIVITY

Divide the class into small groups. Have them imagine a world with no ocean resources. Ask them to brainstorm to come up with a list of activities that would no longer be possible. Examples might include eating certain types of seafood, shipping goods by sea, or deep-sea diving. Have each group share its list with the class and discuss the importance of ocean resources.

Directed Reading Worksheet Section 4

SECTION 4 READING WARM-UP

Terms to Learn

desalination

What You'll Do

- List two methods of harvesting the ocean's living resources.
- List nonliving resources in the ocean.
- Describe the ocean's energy resources.

Resources from the Ocean

The ocean offers a seemingly endless supply of resources. Food, raw materials, energy, and drinkable water are all harvested from the ocean. And there are probably undiscovered resources in unexplored regions of the ocean. As human populations have grown, however, the demand for these resources has increased while the availability has decreased.

Living Resources

People have been harvesting marine plants and animals for food for thousands of years. Many civilizations formed in coastal regions that were rich enough in marine life to support growing human populations. Read on to learn how humans harvest marine life today.

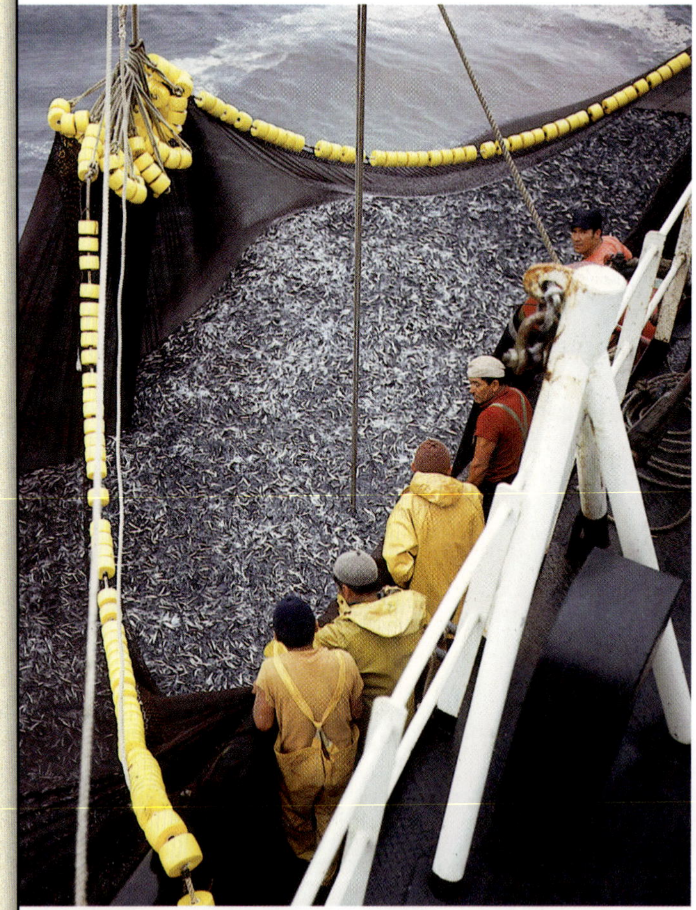

Fishing the Ocean Harvesting food from the ocean is a multibillion-dollar industry. Of all the seafood taken from the ocean, fish are the most abundant. Almost 75 million tons of fish are harvested each year. With improved technology, such as sonar and drift nets, fishermen have become better at locating and taking fish from the ocean. **Figure 15** illustrates how drift nets are used. In recent years, many people have become concerned that we are overfishing the ocean—taking more fish than can be naturally replaced. Also, a few years ago, the public became aware that animals other than fish, especially dolphins and turtles, were accidentally being caught in drift nets. Today the fishing industry is making efforts to prevent overfishing and damage to other wildlife from drift nets.

Figure 15 *Drift nets are fishing nets that cover kilometers of ocean. Fishermen can harvest entire schools of fish in one drift net.*

CONNECT TO ENVIRONMENTAL SCIENCE

Factory trawlers are enormous boats that pull nets as large as four football fields and can harvest 400 tons of fish in a single haul. They can stay at sea for months catching, processing, freezing, and packaging fish. Because of their incredible cost (as much as $40 million), they must catch vast quantities of fish to remain operational.

Factory trawlers threaten marine ecosystems because they can deplete an entire local fish population and quickly move on. They are also criticized for netting large quantities of *bycatch,* unwanted fish and sea-life that are caught and thrown overboard, often dead or dying. Have students find out more about these ships and the environmental controversies that surround them.

Farming the Ocean As overfishing reduces fish populations and laws regulating fishing become stricter, it is becoming more difficult to supply our demand for fish. To compensate for this, many ocean fish, such as salmon and turbot, are being captively bred in fish farms. Fish farming requires several holding ponds, each containing fish at a certain level of development. **Figure 16** shows a holding pond in a fish farm. When the fish are old enough, they are harvested and packaged for shipping.

Figure 16 *Consuming fish raised in a fish farm helps reduce the number of fish harvested from the ocean.*

Savory Seaweed Fish are not the only seafood harvested in a farmlike setting. Shrimp, oysters, crabs, and mussels are raised in enclosed areas near the shore. Mussels and oysters are grown attached to ropes, as shown in **Figure 17.** Huge nets line the nursery area, preventing the animals from being eaten by their natural predators.

Many species of algae, commonly known as seaweed, are also harvested from the ocean. For example, kelp, a seaweed that grows as much as 33 cm a day, is harvested and used as a thickener in jellies, ice cream, and similar products. The next time you enjoy your favorite ice cream, remember that without seaweed, it would be a runny mess! Seaweed is rich in protein, and several species of seaweed are staples of the Japanese diet. For example, the rolled varieties of sushi, a Japanese dish, are wrapped in seaweed.

Figure 17 *In addition to fish, there are many other types of seafood, such as these mussels, that are raised in farms.*

Science Bloopers

Because starfish eat oysters, oyster fishermen have always battled starfish. In this war, when fishermen pulled up a starfish with a clump of oysters, they cut up the starfish and tossed it overboard. Then scientists pointed out that starfish can regenerate from their parts and that by cutting them up, the fishermen were actually making more starfish!

2 Teach

Reading Strategy

Prediction Guide Before students read the passage about living ocean resources, ask them to write an answer for the following question:

What are two problems you think might be associated with fishing the oceans? (Sample answers: overfishing, pollution, accidentally catching other animals in drift nets)

Debate

Fishing the Ocean Lead students in a debate about the costs and benefits of fishing the oceans. Fish are an important source of protein, but overfishing threatens entire ecosystems. Also, fishing is a major source of employment in the world, but many scientists agree that strong restrictions limiting fishing are important. Help students research and debate these issues, and discuss alternatives to traditional fishing, such as fish farming.

Real-World Connection

Writing In the United States, more than 50 percent of the population lives and works within 80 km of the sea, even though coastal areas account for only 11 percent of the nation's land area. Have students write a short story that accurately describes the route that a fish takes from the ocean to the dinner table. Encourage them to focus on the human characters in their story. Ask students to consider how this ocean resource benefits the people who catch, process, ship, sell, and eat it.

2 Teach, continued

Real-World Connection

Have students make a list of all their daily activities that rely on petroleum. Answers might include taking a hot shower, using a hair dryer, cooking breakfast, driving to school, turning on lights, or washing and drying clothes. Allow time for students to share their lists with classmates, then encourage them to brainstorm about ways they might reduce their reliance on fossil fuels.

Group Activity

Writing Divide the class into small groups, and encourage each to select a different ocean resource. Have them work together to write a public service announcement designed to convince the public of the resource's value to people. Have them include ways people can help conserve the resource. Ask students to present their announcements to the class.

Cross-Disciplinary Focus

Language Arts Read selections from Samuel Taylor Coleridge's "The Rime of the Ancient Mariner" in class. Discuss the lines:

> "Water, water, everywhere,
> Nor any drop to drink."

Ask students:

- Where does most drinking water come from? (lakes, rivers, underground)
- How can ocean water be made drinkable? (by desalination)

You may wish to revisit this poem as you discuss ocean pollution in the next section.

Brain Food

Did you know that the United States imports more oil than any other nation? Because oil is a nonrenewable resource, scientists continue to research alternatives to fossil fuels, such as solar energy and wind-generated electricity.

Nonliving Resources

Humans harvest many types of nonliving resources from the ocean. These resources provide raw materials, drinkable water, and energy for our expanding population. Some resources are easily obtained, while others are rare or difficult to harvest.

Oil and Natural Gas Modern civilization continues to be very dependent on oil and natural gas as major sources of energy. Oil and natural gas are *nonrenewable resources,* which means that they are used up faster than they can be replenished naturally. Both oil and natural gas are found under layers of impermeable rock. Petroleum engineers must drill through this rock in order to reach the resources.

Searching for Oil How do petroleum engineers know where to drill for oil and natural gas? Ships with seismic equipment are used for this purpose. Special devices send powerful pulses of sound to the ocean floor. The pulses travel through the water and penetrate the rocks below. The pulses are then reflected back toward the ship, where they are recorded by electronic equipment and analyzed by a computer. The computer readings, such as the one in **Figure 18,** indicate how rock layers are arranged below the ocean floor. Petroleum geologists use these readings to locate a promising area to drill.

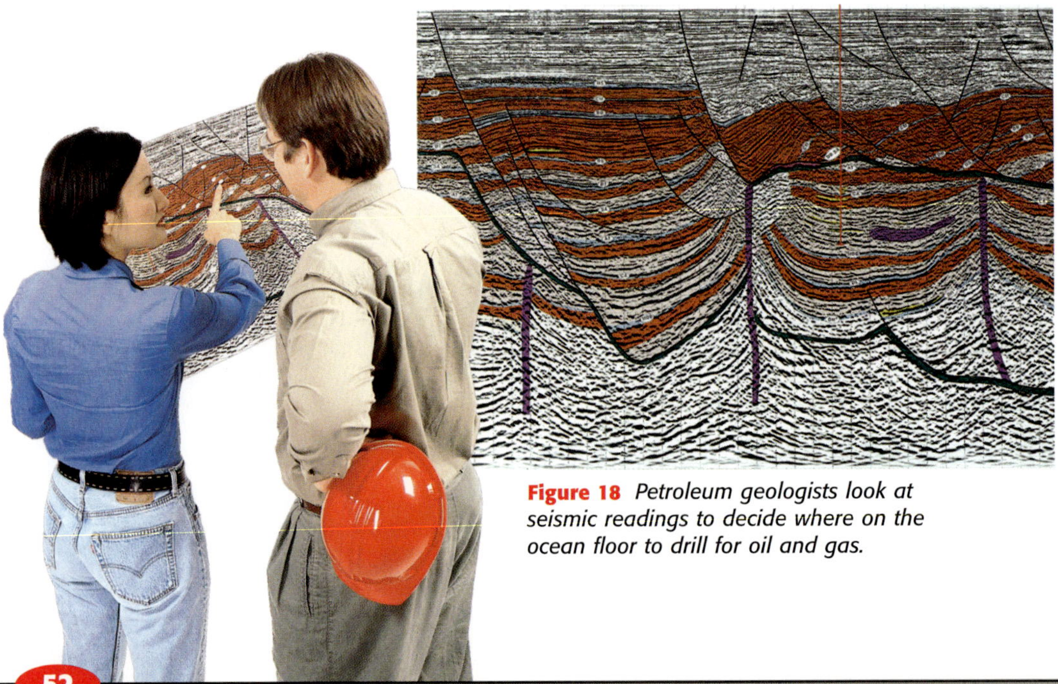

Figure 18 *Petroleum geologists look at seismic readings to decide where on the ocean floor to drill for oil and gas.*

Homework

PORTFOLIO **Research** Help students research the dangers and drawbacks of petroleum drilling and petroleum use. They should look for information about how heavily the world economy depends on petroleum products and why people should be finding cheap, reliable alternative fuels.

Is That a Fact!

Much of the world's petroleum deposits formed millions of years ago when dead organisms accumulated on the ocean floor. Crushed and buried deep beneath the surface by the overlying sediment, this organic matter eventually turned into oil or natural gas.

Fresh Water and Desalination In some areas of the world where fresh water is limited, people desalinate ocean water. **Desalination** is the process of evaporating sea water so that the water and the salt separate. As the water cools and condenses, it is collected and processed for human use. But desalination is not as simple as it sounds, and it is very costly. Countries with an adequate amount of annual rainfall rely on the fresh water provided by precipitation and therefore do not need costly desalination plants. Some countries located in arid regions of the world must build desalination plants to provide an adequate supply of fresh water. Saudi Arabia, located in the desert region of the Middle East, has one of the largest desalination plants in the world.

Sea-Floor Minerals Mining companies are very interested in mineral nodules that are lying on the ocean floor. These nodules are made mostly of manganese, which can be used to make certain types of steel. They also contain iron, copper, nickel, and cobalt. Other nodules are made of phosphates, which are used in making fertilizer.

Nodules are formed from dissolved substances in sea water that stick to solid objects, such as pebbles. As more substances stick to the coated pebble, a nodule begins to grow. Manganese nodules range from the size of a marble to the size of a soccer ball. The photograph in **Figure 19** shows a number of nodules scattered across the ocean floor. It is believed that 15 percent of the ocean floor is covered with these nodules. However, they are located in the deeper parts of the ocean, and mining them is costly and difficult.

QuickLab

How Much Fresh Water Is There?

1. Fill a large **beaker** with 1,000 mL of **water.** This represents all the water on Earth.
2. Carefully pour 970 mL from the beaker into a **graduated cylinder.** This represents the amount of water in the ocean.
3. Pour another 20 mL from the beaker into a **second graduated cylinder.** This represents the amount of water frozen in icecaps and glaciers.
4. Pour another 5 mL into a **third graduated cylinder.** This represents nonconsumable water on land.
5. Take a look at the leftover water. This represents Earth's supply of fresh water.

Put freshwater problems on ice! Turn to page 67 to find out how.

Figure 19 *These manganese nodules could make you wealthy if you knew an affordable way to mine them.*

3 Extend

QuickLab

MATERIALS

For Each Student:
- large beaker
- 3 graduated cylinders

CONNECT TO PHYSICAL SCIENCE

Explain that the mineral nodules shown in **Figure 19** form much like pearls or rock candy—a solid precipitates out of a chemical solution and adheres to a particle. This particle could be nearly anything—a small pebble or even a piece of shell. The nodules grow larger in a process called accretion as more solids precipitate out of the solution. Students could model accretion by making rock candy with a supersaturated sugar solution and a piece of string.

Homework

Graphing Help students use references to find out how much of the world's energy needs are met by the following:

oil, natural gas, coal, tidal energy, geothermal energy, wave energy, hydroelectric energy, solar energy, and wind energy

Have students prepare pie charts or bar graphs of their findings.

Reinforcement Worksheet
"The Oceans and Us"

WEIRD SCIENCE

Buried in sea floor sediments at depths greater than 500 m below sea level, an ice called *methane hydrate* may be twice as abundant as all other fossil fuels combined. Methane hydrate has also been discovered in the permafrost of Siberia and Alaska. Clean burning and abundant, methane hydrate could be the fuel of the future, but engineers have not yet discovered an efficient way to mine it. Have students find out about some of the interesting proposals to mine this resource and also about the role decomposing methane hydrates may play in global warming.

4 Close

Quiz

1. Explain some of the ecological risks of using drift nets. (Using drift nets can lead to overfishing, which depletes fish populations. Drift nets also accidentally trap other marine animals such as dolphins and turtles.)
2. What are nonrenewable resources? Give two examples. (Nonrenewable resources are resources that are used up faster than they can be replenished naturally. Natural gas and oil are nonrenewable resources.)
3. What is desalination, and why is it done? (Desalination is the process of evaporating sea water so that the water and the salt separate. Desalination provides fresh water for human use.)

Alternative Assessment

Challenge students to make drawings that demonstrate how tidal energy is harnessed. Encourage them to explain the process to the class, using their drawings as visual aids. Their drawings and explanations should reflect the understanding that water must rush through a narrow channel and move turbines to generate electricity.

Figure 20 **Using Tides to Generate Electricity**

As the tide rises, water enters a bay behind a dam. The gate then closes at high tide.

The gate remains closed as the tide lowers.

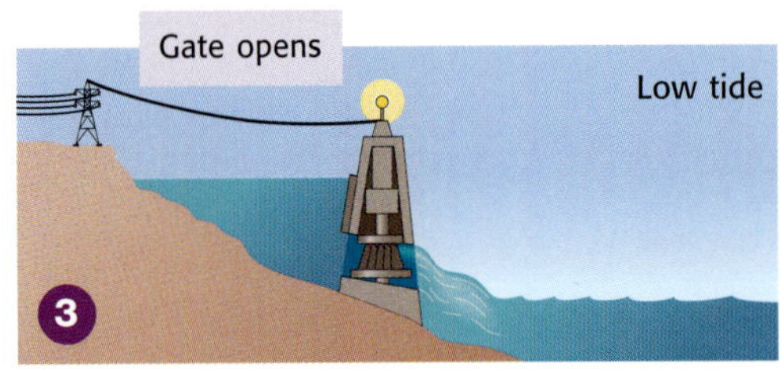

At low tide, the gate opens, and the water rushes through the dam, moving the turbines, which in turn create electricity.

Tidal Energy The ocean creates several types of energy resources simply because of its constant movement. The gravitational pull of the sun and moon causes the ocean to rise and fall as tides. *Tidal energy,* energy generated from the movement of tides, is an excellent alternative source of energy. If the water during high tide can be rushed through a narrow coastal passageway, the water's force can be powerful enough to generate electricity. **Figure 20** shows how this works. Tidal energy is a clean, inexpensive, and renewable resource once the dam is built. A *renewable resource* can be replenished, in time, after being used. Unfortunately, tidal energy is practical only in a few areas of the world, where the coastline has shallow, narrow channels. For example, the coastline at Cook Inlet, in Alaska, is perfect for generating tidal power.

Wave Energy Have you ever stood on the beach and watched as waves crashed onto the shore? This constant motion is an energy resource. Wave energy, like tidal energy, is a clean, renewable resource.

Recently, computer programs have been developed to analyze the energy of waves. Researchers have located certain areas of the world where wave energy can generate enough electricity to make it worthwhile to build power plants. Wave energy in the North Sea is strong enough to produce power for parts of Scotland and England.

internetconnect

SCILINKS NSTA
TOPIC: Ocean Resources
GO TO: www.scilinks.org
***sci*LINKS NUMBER:** HSTE320

SECTION REVIEW

1. List two methods of harvesting the ocean's living resources.
2. Name four nonliving resources in the ocean.
3. **Interpreting Graphics** Take another look at Figure 20. As the tide is rising, will the gate be open or closed? How might this affect the turbines?

54

Critical Thinking Worksheet "Chain Reaction"

internetconnect

SCILINKS NSTA
TOPIC: Ocean Resources
GO TO: www.scilinks.org
***sci*LINKS NUMBER:** HSTE320

Answers to Section Review

1. Answers will vary. Two methods of harvesting the ocean's resources are fishing and farming the ocean.
2. Any four of the following nonliving resources are acceptable: oil, natural gas, fresh water, minerals, tidal energy, and wave energy.
3. As the tide is rising, the gate will be open. The turbines may turn while the water passes through the dam from the open ocean.

SECTION 5
READING WARM-UP

Terms to Learn

nonpoint-source pollution

What You'll Do

- List different types of ocean pollution.
- Explain how to prevent or minimize different types of ocean pollution.
- Outline what is being done to control ocean pollution.

Ocean Pollution

Humans have used the ocean for waste disposal for hundreds, if not thousands, of years. This has harmed the organisms that live in the oceans as well as animals that depend on marine organisms. People are also affected by polluted oceans. Fortunately, we are becoming more aware of ocean pollution, and we are learning from our mistakes.

Sources of Ocean Pollution

There are many sources of ocean pollution. Some of these sources are easily identified, but others are more difficult to pinpoint. Read on to find out where different types of ocean pollution come from and how they affect the ocean.

Trash Dumping People dump trash in many places, including the ocean. In the 1980s, scientists became alarmed by the kind of trash that was washing up on beaches. Bandages, vials of blood, and syringes (needles) were found among the waste. Some of the blood in the vials even contained the AIDS virus. The Environmental Protection Agency (EPA) began an investigation and discovered that hospitals in the United States produce an average of 3 million tons of medical waste each year. And where does some of this trash end up? You guessed it—in the ocean. Because of stricter laws, much of this medical waste is now buried in sanitary landfills. However, dumping trash in the deeper part of the ocean is still a common practice in many countries.

Figure 21 *This barge is headed out to the open ocean, where it will dump the trash it carries.*

55

SECTION 5

Focus

Ocean Pollution

In this section, students learn about the sources and effects of ocean pollution. They also learn about the strategies that are used to minimize ocean pollution. The section examines the effects and cleanup of a major oil spill, and discusses national and international action to reduce ocean pollution.

Bellringer

Ask students to write a few sentences about how ocean pollution could affect their lives. Then have students list the ways they contribute to ocean pollution in their daily lives. Ask groups to brainstorm about the ways they could reduce ocean pollution.

1 Motivate

Activity

Cleaning Up an Oil Spill Give each pair of students a pan of water. Pour about 5 mL of vegetable oil into each pan. Have students think of ways to remove the oil from the pan *without pouring out the water.* Students can experiment with methods such as absorbing the oil with a paper towel or scooping it out with a spoon. Discuss the results of students' efforts. Tell them that this problem is similar to dealing with a petroleum spill in the ocean. Ask students what other factors would be important to containing an oil spill in the ocean. (wind, currents, waves, the size of the spill, the type of oil, the temperature of the ocean water)

SCIENCE HUMOR

Huntsville, Alabama, hairdresser Phil McCrory has patented a way to use discarded human hair to clean up oil spills. He tested the idea in his son's wading pool using a pair of pantyhose filled with hair. The idea works so well that McCrory has attracted the attention of NASA!

internetconnect

sciLINKS NSTA

TOPIC: Ocean Pollution
GO TO: www.scilinks.org
sciLINKS NUMBER: HSTE323

Directed Reading Worksheet Section 5

2 Teach

Meeting Individual Needs

Learners Having Difficulty
Draw students' attention to the photographs of nonpoint-source pollutants in **Figure 22.** Ask them to think about their daily lives and to list ways they might be contributing to nonpoint-source pollution. Ask them to help you list ways that they could reduce their contribution to nonpoint-source pollution. Sheltered English

Group Activity

Have student groups report on the current status of the communities and ecosystems of Prince William Sound in the wake of the *Exxon Valdez* oil spill. Do fishing communities feel that they have been compensated?

Homework

Investigate Your Area Inform students that many companies are creating products that are marketed to be less harmful to the environment. These claims have varying degrees of truth. Have groups choose a product and investigate the environmental statements that are used to market it. For example, a group could visit the detergent aisle of a grocery store to compare products. Have them prepare a list of products claiming to be environmentally friendly and their primary ingredients. Encourage students to use Internet or library resources to determine whether the ingredients listed could be harmful to the environment. Another group could compare the percentage of post-consumer waste in recycled paper products. Have groups present their findings to the class.

Environment Connection

In Austin, Texas, sludge is used to make a compost called *Dillo Dirt.* (*Dillo* refers to *armadillo,* a small aardvark-like mammal common in Texas.) Instead of polluting the Gulf of Mexico, Austinites are using sludge to grow beautiful and beneficial gardens.

Sludge Dumping By 1990, the United States alone had discharged 38 trillion liters of treated sludge into the waters along its coasts. What is sludge, and why is it so bad? To answer this question, we need to define *raw sewage.*

Raw sewage is all the liquid and solid wastes that are flushed down toilets and poured down drains. After collecting in sewer drains, raw sewage is sent through a treatment plant, where it undergoes a cleaning process that removes the solid waste. The solid waste is called *sludge.* In many areas, people dump sludge into the ocean several kilometers offshore, intending for it to settle to and stay on the ocean floor. Unfortunately, currents sometimes stir the sludge up and move it closer to shore. This can pollute beaches and kill marine life. Many countries have banned sludge dumping, but it continues to occur in many areas of the world.

Nonpoint-Source Pollution We usually think of water pollution as coming from large factories, but you may be surprised to know that most of the pollution comes from everyday citizens doing everyday things. This type of pollution, which is shown in **Figure 22,** is called **nonpoint-source pollution** because you cannot pinpoint its exact source. How does this pollution get into the ocean? All waste water and runoff eventually enter a body of water, usually a stream. Every stream leads to a river, and every river leads to the ocean.

Figure 22 *Nonpoint-source pollution contributes significantly to ocean pollution. What can you do to cut down on nonpoint-source pollution?*

56

Is That a Fact!

The world's first major oil spill occurred on March 18, 1967. The tanker *Torrey Canyon* ran aground off the coast near Cornwall, England. The ship spilled about 870,000 barrels of oil, more than three times the oil spilled by the *Exxon Valdez.*

Oil Spills Because oil is in such high demand across the world, large tankers must transport billions of barrels of it across the oceans. If not handled properly, these transports can quickly turn disastrous. In 1989, the supertanker *Exxon Valdez* struck a reef and spilled more than 260,000 barrels of crude oil. The effect of this accident on wildlife was catastrophic. Many animals died. Alaskans who made their living from fishing lost their businesses. Although many animals were saved, as shown in **Figure 23**, and the Exxon Oil Company spent $2.5 billion to try to clean up the mess, Alaska's wildlife and economy will continue to suffer for decades.

Today many oil companies are using new technology to safeguard against oil spills. Tankers are now being built with two hulls instead of one. This prevents oil from spilling into the ocean if the outside hull of the ship is damaged. **Figure 24** illustrates the design of a double-hulled tanker.

Figure 23 *Many oil-covered animals were rescued and cleaned after the* Exxon Valdez *spill.*

BRAIN FOOD

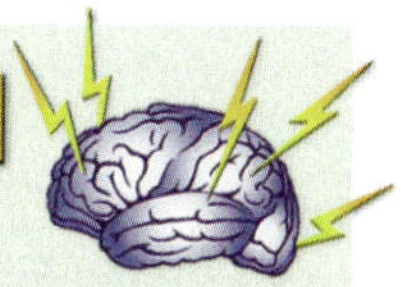

Within the first few weeks of the *Exxon Valdez* oil spill, more than half a million birds, including 109 endangered bald eagles, were covered with oil and drowned. Almost half the sea otters in the area also died either from drowning or from being poisoned by the oil.

Figure 24 *If the outside hull of a double-hulled tanker is punctured, the oil will still be contained within the inside hull.*

MATH and MORE

Challenge students to use library resources to find out which nations contribute the most to ocean pollution. Have them construct pie graphs to show their results and write a short essay explaining the trends they observe.

DISCUSSION

Nuclear Waste and the Ocean
Remind students that the ocean tends to separate into layers because temperature and salinity differences create layers with different densities.

Point out that the world community is confronted with the dilemma of nuclear waste disposal. One proposed solution is to bury weighted barrels of waste in the sediment of the ocean floor. Encourage students to consider the implications of such a strategy. As nuclear waste decays, it releases energy. What would happen if water contaminated by nuclear waste heated the deep zone?

COOPERATIVE LEARNING

Divide the class into small groups. Ask each group to imagine that they are the campaign managers for a candidate trying to become the city manager of a coastal town. Tell them that they must convince voters that they are concerned for the environmental health of the ocean and have a realistic plan to preserve the beaches and marine life. Have them write a campaign speech outlining the sources of pollution and the solutions they propose. Ask one student from each group to present the speech to the class. Allow the class to vote for the most convincing candidate.

CONNECT TO ENVIRONMENTAL SCIENCE

The Oil Pollution Act of 1990 was a direct response to the *Exxon Valdez* oil spill. The controversial bill had been debated for 14 years; it passed swiftly in the aftermath of the disaster. Under the law, all oil tankers operating in United States waters must be double-hulled by 2015. Compliance has been slow, however; many oil companies have been reluctant to replace the aging boats in their fleets with hundred-million-dollar double-hulled ships. As of 1999, of the 3,294 oil tankers operating worldwide, only 876 were double-hulled.

3 Extend

LabBook PG 104
Investigating an Oil Spill

Activity

Divide the class into small groups. Provide each group with poster board and markers. Direct each group to work together to create a poster designed to educate the public about the need for clean water. Ask them to focus on ways people can minimize the pollution of our oceans. Consider displaying the posters around the school to educate students and teachers. As an extension, have students draft a letter to a United States Congress member that outlines each group's ideas. Sheltered English

Research

Help students find information to investigate the environmental impact of one of the following:

developing areas where salt marshes once were, draining mangrove swamps for development, the pollution of ocean waters with agricultural runoff, damming rivers and streams that flow into the ocean, using dynamite to catch fish (common in the Pacific and particularly harmful to coral reefs), or destroying coral reefs to obtain construction materials

Encourage students to focus on the effects of these human behaviors on other organisms, and challenge them to suggest alternatives

Saving Our Ocean Resources

Although humans have done much to harm the ocean's resources, we have also begun to do more to save them. From international treaties to volunteer cleanups, efforts to conserve the ocean's resources are making an impact around the world.

Nations Take Notice When ocean pollution reached an all-time high, many countries recognized the need to work together to solve the problem. In 1989, 64 countries ratified a treaty that prohibits the dumping of mercury, cadmium compounds, certain plastics, oil, and high-level radioactive wastes into the ocean. Many other international agreements restricting ocean pollution have been made, but enforcing them is often difficult.

In spite of efforts to protect the ocean, waste dumping and oil spills still occur, and contaminated organisms continue to wash ashore. Why are the laws not working as well as they should? Enforcing these laws takes money and human resources, and many agencies are lacking in both.

LabBook

Why worry about a few drops of oil? You might be surprised that a little goes a long way. Turn to page 104 in the LabBook to learn more.

Action in the United States The United States, like many other countries, has taken additional measures to control local pollution. In 1972, Congress passed the Clean Water Act, which put the EPA in charge of issuing permits for any dumping of trash into the ocean. Later that year, a stricter law was passed. The U.S. Marine Protection, Research, and Sanctuaries Act prohibits the dumping of any material that would affect human health or welfare, the marine environment or ecosystems, or businesses that depend on the ocean.

APPLY

Ocean Treaty

Get together with your classmates and divide yourselves into three groups: Nation A, Nation B, and Nation C. All three nations are located near the ocean, and all three nations share borders. Nation A has a very rich supply of oil, which it transports around the world. Nation B currently depends on nuclear energy and has many nuclear power plants near its shores. Nation B has no place on land to store radioactive waste from its nuclear power plants. Nation C sells nuclear technology to Nation B, buys oil from Nation A, and has the world's most diverse coastal ecosystem. The three nations must form a treaty to safeguard against ocean pollution without seriously harming any of their economies. Can you do it?

Answers to APPLY

Although there are no specific answers, students should consider the following factors:

1. Nation A and Nation C must keep each other satisfied because they depend on the sale of oil from Nation A to Nation C.
2. Nation B and Nation C must keep each other satisfied because Nation B buys nuclear technology from Nation C.
3. Nation C must protect its coastal ecosystem, but Nation C also depends on Nation A's transport of oil and Nation B's use of nuclear energy, both of which can potentially threaten the coastal ecosystem.

Figure 25 *The Adopt-a-Beach program in Texas has been a huge success.*

Citizens Take Charge Citizens of many countries have demanded that their governments do more to solve the growing problem of ocean pollution. Because of public outcry, the United States now spends more than $130 million each year monitoring the oceans. United States citizens have also begun to take the matter into their own hands. In the early 1980s, citizens began organizing beach cleanups. One of the largest cleanups is the semiannual Adopt-a-Beach program, shown in **Figure 25,** that originated with the Texas Coastal Cleanup campaign. Millions of tons of trash have been gathered from the beaches, and people are being educated about the hazards of ocean dumping.

Though governments pass laws against ocean dumping, keeping the oceans clean is everyone's responsibility. The next time you and your family visit the beach, make sure the only items you leave behind on the sand are hermit crabs, shells, and maybe a few sand dollars.

SECTION REVIEW

1. List three types of ocean pollution. How can each of these types be prevented or minimized?
2. Which type of ocean pollution is most common?
3. **Summarizing Data** List and describe three measures that governments have taken to control ocean pollution.

internetconnect

SCI LINKS NSTA

TOPIC: Ocean Pollution
GO TO: www.scilinks.org
***sci*LINKS NUMBER:** HSTE323

Answers to Section Review

1. Any of the following are acceptable: trash dumping, sludge dumping, nonpoint-source pollution, and oil spills. Trash can be dumped in sanitary landfills, sludge can be used to make compost, people can reduce nonpoint-source pollution by changing daily behaviors. Tankers with double hulls can be used to transport oil. All of these types of pollution can be prevented or minimized through legislation, education, and actions by citizens.
2. nonpoint-source pollution
3. Three measures taken by governments to control ocean pollution are described on page 58.

DEBATE

Clean Water Legislation

Challenge groups of students to research legislation intended to control ocean pollution. Students might learn more about the Clean Water Act, the MARPOL Act, or the U.S. Marine Protection, Research, and Sanctuaries Act, all of which limit dumping. Groups should debate the pros and cons of each bill.

4 Close

Quiz

1. Why do oil spills pose a long-term risk? (Organisms living in the ocean are part of a complex web of relationships. If one group of organisms are affected, many other groups could be threatened. In addition, the oil tends to concentrate in the tissues of living organisms and can persist long after the spill has been cleaned up.)
2. Why is sludge dumping a threat to marine and human life? (Sludge dumping can cause diseases, kill marine organisms, and pollute beaches.)

ALTERNATIVE ASSESSMENT

Concept Mapping Have students recall the sources of ocean pollution discussed in this section, the effects of pollution on oceans and wildlife, and the actions being taken by nations and individuals to limit and reduce pollution. Ask them to prepare a concept map that organizes and compares this information. Challenge students to propose additional suggestions for pollution prevention.

Probing the Depths
Teacher's Notes

Time Required

One 45-minute class period

Lab Ratings

EASY → HARD

- TEACHER PREP: 1
- STUDENT SET-UP: 2
- CONCEPT LEVEL: 2
- CLEAN UP: 1

MATERIALS

The materials listed on the student page are enough for each student or for a group of 2–4 students.

Safety Caution

Remind students to review all safety cautions and icons before beginning this lab activity.

Preparation Notes

You may wish to ask students to provide their own shoe boxes. It may be safer and easier if you punch the holes along the center of the lids for the students prior to class. You may use corrugated cardboard instead of modeling clay. If you do, shape the cardboard into steps and ridges to model changes in depth. With this type of model, students should use the eraser end of an unsharpened pencil to measure depths.

Making Models Lab

Probing the Depths

In the 1870s, the crew of the ship the HMS *Challenger* used a wire and a weight to discover and map some of the deepest places in the world's oceans. The crew members tied a wire to a weight and dropped the weight overboard. When the weight reached the bottom of the ocean, they hauled the weight back up to the surface and measured the length of the wet wire. In this way, they were eventually able to map the ocean floor. In this activity, you will model this method of mapping by making a map of an ocean-floor model.

MATERIALS

- modeling clay
- shoe box and lid
- scissors
- 8 unsharpened pencils
- metric ruler

Procedure

1. Use the clay to make a model ocean floor in the shoe box. Give the floor some mountains and valleys.
2. Cut eight holes in a line along the center of the lid. The holes should be just big enough to slide a pencil through. Close the box.
3. Exchange boxes with another student or group of students. Do not look into the box.
4. Copy the table on the next page into your ScienceLog. Also make a copy of the graph on the next page.
5. Measure the length of the probe (pencil) in centimeters. Record the length in your data table.
6. Gently insert the probe into the first hole position in the box until it touches the bottom. Do not push the probe down; pushing it down could affect your reading.
7. Making sure the probe is straight up and down, measure the length of probe showing above the lid. Record your data in the data table.
8. Use the following formula to calculate the depth in centimeters:

$$\left(\textit{original length of probe}\right) - \left(\textit{amount of probe showing}\right) = \textit{depth in cm}$$

60

CLASSROOM TESTED & APPROVED

Tracy Jahn
Berkshire Jr.-Sr. High School
Canaan, New York

Ocean Depth Chart

Hole position	Original length of probe (cm)	Amount of probe showing (cm)	Depth in centimeters	Depth in meters (cm × 200)
1				
2				
3				
4				
5				
6				
7				
8				

9. Use the scale 1 cm = 200 m to convert the depth in centimeters to meters to better represent real ocean depths. Add the data to your table.

10. Transfer the data to your graph for position 1.

11. Repeat steps 5–10 for the additional positions in the box.

12. After plotting all the points on your graph, connect the points with a smooth curve.

13. Put a pencil in each of the holes in the shoe box. Compare the rise and fall of the set of pencils with your graph.

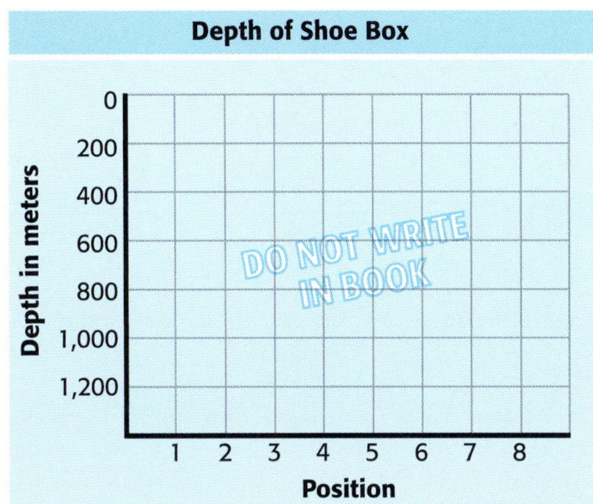

Analysis

14. How deep was your deepest point? your shallowest point?

15. Did your graph resemble the ocean-floor model, as shown by the pencils? If not, why not?

16. What difficulties might scientists have when measuring the real ocean floor? Do they ever get to "open the box"? Explain your answer.

Answers

14. Answers will vary.

15. Answers will vary. Accept all reasonable responses. Discrepancies between the model and the graph may be attributed to incorrectly plotting the points, failing to keep the probe vertical, or pushing the probe into the clay.

16. Scientists may encounter many difficulties in measuring the ocean floor. Students may note that the extreme depths of the ocean make it difficult to map the ocean floor accurately. Students may also note that scientists are often unable to "open the box" to check their measurements; that is, the ocean is too deep in places and too vast to fully explore.

Datasheets for LabBook

Science Skills Worksheet "Introduction to Graphs"

Chapter Highlights

Vocabulary Definitions

Section 1

salinity a measure of the amount of dissolved salts and other solids in a given amount of liquid

thermocline a layer of ocean water extending from 300 m below sea level to about 700 m below sea level in which water temperature drops with increased depth faster than in other layers of the ocean

water cycle the continuous movement of water from water sources into the air, onto land, into and over the ground, and back to the water sources; a cycle that links all of the Earth's solid, liquid, and gaseous water together

Section 2

continental shelf the flattest part of the continental margin

continental slope the steepest part of the continental margin

continental rise the base of the continental slope

abyssal plain the broad, flat portion of the deep-ocean basin

mid-ocean ridge a long mountain chain that forms on the ocean floor where tectonic plates pull apart; usually extends along the center of ocean basins

rift valley a valley that forms in a rift zone between diverging tectonic plates

seamount an individual mountain of volcanic material on the abyssal plain

ocean trench a seemingly bottomless crevice in the deep-ocean basin that forms where one oceanic plate is forced underneath a continental plate or another oceanic plate

Chapter Highlights

SECTION 1

Vocabulary

salinity *(p. 36)*
thermocline *(p. 37)*
water cycle *(p. 39)*

Section Notes

- The four oceans as we know them today formed within the last 300 million years.
- Salts have been added to the ocean for billions of years.
- The three temperature zones of ocean water are the surface zone, thermocline, and deep zone.
- The ocean plays the largest role in the water cycle.
- The ocean stabilizes Earth's conditions by absorbing and retaining thermal energy.

SECTION 2

Vocabulary

continental shelf *(p. 42)*
continental slope *(p. 42)*
continental rise *(p. 42)*
abyssal plain *(p. 42)*
mid-ocean ridge *(p. 43)*
rift valley *(p. 43)*
seamount *(p. 43)*
ocean trench *(p. 43)*

Section Notes

- The ocean floor is divided into zones based on depth and slope.
- The continental margin consists of the continental shelf, the continental slope, and the continental rise.
- The deep-ocean basin consists of the abyssal plain, with features such as mid-ocean ridges, rift valleys, seamounts, and ocean trenches.
- In addition to directly studying the ocean floor, scientists indirectly study the ocean floor using sonar and satellites.

Skills Check

Math Concepts

PERCENTAGES Percentages are a way of describing the parts within a whole. Percentages are expressed in hundredths. Take another look at Figure 1 on page 36. The pie chart shows the percentages of dissolved solids in ocean water. The amount of chlorine (Cl) dissolved in the ocean is 55 percent. This means that 55 of every 100 parts of dissolved solids in the ocean are chlorine.

Visual Understanding

TEMPERATURE ZONES Look back at the line graph on page 37 to review why the temperature of the ocean decreases with increasing depth.

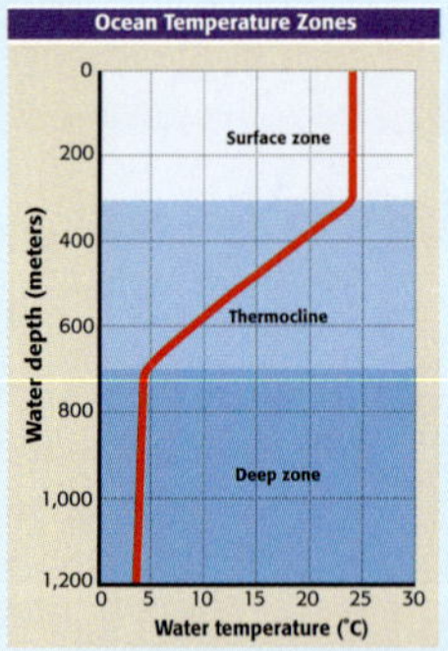

Lab and Activity Highlights

Probing the Depths PG 60

Investigating an Oil Spill PG 104

Datasheets for LabBook (blackline masters for these labs)

SECTION 3

Vocabulary

plankton *(p. 46)*
nekton *(p. 46)*
benthos *(p. 46)*
benthic environment *(p. 47)*
pelagic environment *(p. 49)*

Section Notes

- There are three main groups of marine life—plankton, nekton, and benthos.
- The two main ocean environments—the benthic and pelagic environments—are divided into ecological zones based on the locations of organisms that live in the environments.

SECTION 4

Vocabulary

desalination *(p. 53)*

Section Notes

- Humans depend on the ocean for living and nonliving resources.
- Ocean farms raise fish and other marine life to help feed growing human populations.
- Nonliving ocean resources include oil and natural gas, fresh water, minerals, and tidal and wave energy.

SECTION 5

Vocabulary

nonpoint-source pollution *(p. 56)*

Section Notes

- Types of ocean pollution include trash dumping, sludge dumping, nonpoint-source pollution, and oil spills.
- Nonpoint-source pollution cannot be traced to specific points of origin.
- Efforts to save ocean resources include international treaties and volunteer cleanups.

Labs

Investigating an Oil Spill *(p. 104)*

internetconnect

go.hrw.com **GO TO:** go.hrw.com

Visit the HRW Web site for a variety of learning tools related to this chapter. Just type in the keyword:

KEYWORD: HSTOCE

sciLINKS NSTA **GO TO:** www.scilinks.org

Visit the **National Science Teachers Association** on-line Web site for Internet resources related to this chapter. Just type in the *sci*LINKS number for more information about the topic:

TOPIC: Exploring Earth's Oceans **sciLINKS NUMBER:** HSTE305
TOPIC: The Ocean Floor **sciLINKS NUMBER:** HSTE310
TOPIC: Life in the Oceans **sciLINKS NUMBER:** HSTE315
TOPIC: Ocean Resources **sciLINKS NUMBER:** HSTE320
TOPIC: Ocean Pollution **sciLINKS NUMBER:** HSTE323

63

VOCABULARY DEFINITIONS, *continued*

SECTION 3

plankton microscopic organisms that float at or near the ocean's surface

nekton free-swimming organisms of the ocean

benthos organisms that live on or in the ocean floor

benthic environment the ocean floor and all the organisms that live on or in it; also known as the bottom environment

pelagic environment the entire volume of water in the ocean and the marine organisms that live above the ocean floor; also known as the water environment

SECTION 4

desalination the process of evaporating sea water so that the water and the salt separate

SECTION 5

nonpoint-source pollution pollution that comes from many sources and that cannot be traced to specific sites

Vocabulary Review Worksheet

Blackline masters of these Chapter Highlights can be found in the **Study Guide.**

Lab and Activity Highlights

LabBank

Inquiry Labs, Surf's Up!

Whiz-Bang Demonstrations, Fowl Play

EcoLabs & Field Activities, Operation Oil-Spill Cleanup

Calculator-Based Labs, Ocean Floor Mapping

Long-Term Projects & Research Ideas, Your Very Own Underwater Theme Park

Interactive Explorations CD-ROM

CD 2, Exploration 2, "Sea Sick"

Chapter Review Answers

Using Vocabulary

1. continental shelf
2. thermocline
3. Plankton
4. *Desalination* is the process of evaporating sea water so that the water and salt separate.
5. Types of *benthos* include sea stars and clams.
6. An ocean trench forms where one oceanic plate is forced underneath a continental plate or another oceanic plate, while a rift valley forms in a rift zone, where tectonic plates pull apart.
7. Salinity is a measure of the amount of dissolved solids in a given amount of liquid, while desalination is the process of evaporating sea water so that the water and the salt separate.
8. Nekton are free-swimming marine organisms, while benthos are marine organisms that live on or in the ocean floor.
9. The pelagic environment is made up of the water in the ocean and all the organisms that live in the water, while the benthic environment is made up of the ocean floor and all the organisms that live on or in the ocean floor.

Understanding Concepts

Multiple Choice

10. b
11. c
12. d
13. c
14. a
15. a

Chapter Review

USING VOCABULARY

To complete the following sentences, choose the correct term from each pair of terms listed below:

1. The region of the ocean floor that is closest to the shoreline is the __?__. *(continental shelf* or *continental slope)*
2. Below the surface layer of the ocean is a layer of water that gets colder with depth and extends to a depth of 700 m. This layer is called the __?__. *(thermocline* or *benthic environment)*
3. __?__ typically float at or near the ocean's surface. *(Plankton* or *Nekton)*

Correct the wrong terminology in each of the following sentences. A word bank is provided.

4. The water cycle is the process of evaporating sea water so that the water and salt separate.
5. Types of nekton include sea stars and clams.

Word bank: nonpoint-source pollution, plankton, desalination, benthos

Explain the difference between the words in each of the following pairs:

6. ocean trench/rift valley
7. salinity/desalination
8. nekton/benthos
9. pelagic environment/benthic environment

UNDERSTANDING CONCEPTS

Multiple Choice

10. The largest ocean is the
    a. Indian Ocean.
    b. Pacific Ocean.
    c. Atlantic Ocean.
    d. Arctic Ocean.

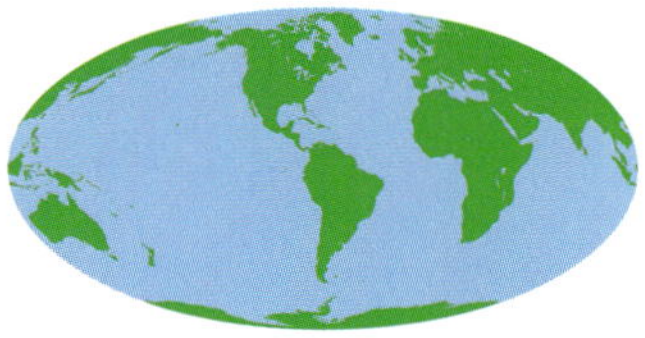

11. One of the most abundant elements in the ocean is
    a. potassium.
    b. calcium.
    c. chlorine.
    d. magnesium.
12. Which of the following affects the ocean's salinity?
    a. fresh water added by rivers
    b. currents
    c. evaporation
    d. all of the above
13. Most precipitation falls
    a. on land.
    b. into lakes and rivers.
    c. into the ocean.
    d. in rain forests.
14. Which benthic zone has a depth range between 200 m and 4,000 m?
    a. bathyal zone
    b. abyssal zone
    c. hadal zone
    d. sublittoral zone
15. The ocean floor and all the organisms that live on it or in it is the
    a. benthic environment.
    b. pelagic environment.
    c. neritic zone.
    d. oceanic zone.

Short Answer

16. Why does coastal water in areas with hotter, drier climates typically have a higher salinity than coastal water in cooler, more humid areas?

17. What is the difference between the abyssal plain and the abyssal zone?

18. How do the continental shelf, the continental slope, the continental rise, and the continental margin relate to each other?

Concept Mapping

19. Use the following term to create a concept map: marine life, plankton, nekton, benthos, benthic environment, pelagic environment.

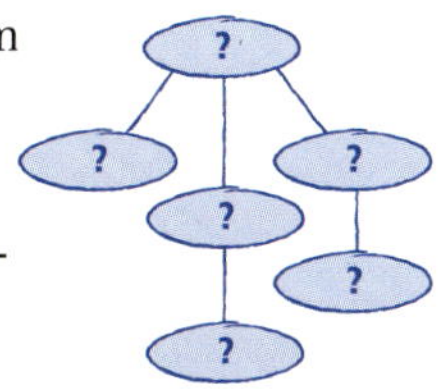

CRITICAL THINKING AND PROBLEM SOLVING

Write one or two sentences to answer the following questions:

20. Other than obtaining fresh water, what benefit comes from desalination?

21. Explain the difference between a bathymetric profile and a seismic reading.

MATH IN SCIENCE

22. Imagine that you are in the kelp-farming business and that your kelp grows 33 cm per day. You begin harvesting when your plants are 50 cm tall. During the first seven days of harvest, you cut 10 cm off the top of your kelp plants each day. How tall would your kelp plants be after the seventh day of harvesting?

INTERPRETING GRAPHICS

Examine the image below, and answer the questions that follow:

Ecological Zones of the Ocean

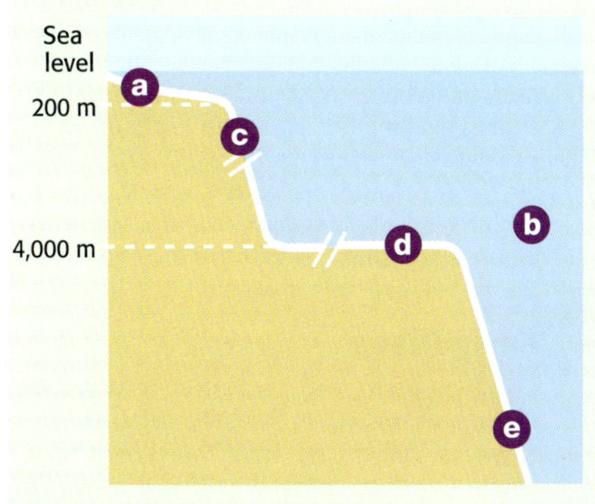

23. At which point *(a, b, c, d, or e)* would you most likely find an anglerfish?

24. At which point would you most likely find tube worms?

25. Which ecological zone is at point *c*? Which depth zone is at point *c*?

26. Name a type of organism you might find at point *e*.

Reading Check-up

Take a minute to review your answers to the Pre-Reading Questions found at the bottom of page 32. Have your answers changed? If necessary, revise your answers based on what you have learned since you began this chapter.

Short Answer

16. because less fresh water runs into the ocean in drier areas and because heat increases the evaporation rate

17. The abyssal plain is the flattest and deepest part of the ocean floor. The abyssal zone is an ecological zone that consists of the abyssal plain and all the organisms that live on or in it.

18. The continental shelf, the continental slope, and the continental rise are sections of the continental margin. Also, the continental rise is part of the continental slope.

Concept Mapping

19. 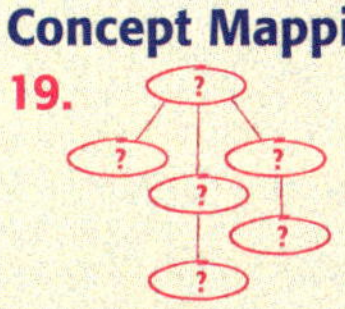 An answer to this exercise can be found in the front of this book.

Critical Thinking and Problem Solving

20. Salt is left behind. Salt can be used for many things.

21. A bathymetric profile shows what the contour of the ocean floor looks like, while a seismic reading shows what rocks beneath the ocean floor look like. (Seismic readings can show the contour of the ocean floor too.)

Math in Science

22. 50 cm + 7(33 cm − 10 cm) = 211 cm

Interpreting Graphics

23. *b*

24. *d*

25. ecological zone = bathyal zone, depth zone = continental slope

26. Accept any one of the following: clams, sponges, or worms.

Concept Mapping Transparency 13

Blackline masters of this Chapter Review can be found in the **Study Guide.**

Teaching Strategy

In addition to being an inventor and an explorer, Jacques Cousteau was a commentator on the twentieth century and its environmental problems. He was criticized as being a populist because he brought information to the public in common language instead of in academic terms. Help the class find information about his condemnation of French nuclear testing or his controversial opinions on overpopulation and animal testing.

You may want to show your students some of Cousteau's inspired documentaries. Some of his full-length features include *The Silent World, World Without Sun,* and *Voyage to the Edge of the World.* Tapes of his many television series *(The Undersea World of Jacques Cousteau, Cousteau Odyssey,* and *Cousteau Amazon)* may also be available at a public library.

ACROSS THE SCIENCES

EARTH SCIENCE • LIFE SCIENCE

Exploring Ocean Life

Jacques Cousteau, born in France in 1910, opened the eyes of countless people to the sea. During his long life, Cousteau explored Earth's oceans and documented the amazing variety of life they contained. Jacques Cousteau was an explorer, environmentalist, inventor, and teacher who inspired millions with his joy and wonder at the watery part of our planet.

Early Dives

Cousteau performed his first underwater diving mission at age 10. At summer camp he was asked to collect trash from the camp's lake. The young Cousteau quickly realized that working underwater without goggles or breathing equipment was a tremendous challenge.

Cousteau had another early underwater experience when he visited Southeast Asia. He saw people diving into the water to catch fish with their bare hands. This fascinated Cousteau. Even at a young age, he was thinking about how to make equipment that would let a person breathe underwater.

▲ *Cousteau in front of the* Calypso II

Underwater Flight

As a young man, Cousteau and some friends developed the aqualung, a self-contained breathing system for underwater exploration. As someone who had often dreamed of flying, Cousteau was thrilled with his invention. After one of his first dives, Cousteau explained, "I experimented with all possible maneuvers—loops, somersaults, and barrel rolls. . . . Delivered from gravity and buoyancy, I flew around in space."

Using the aqualung and other underwater equipment he developed, Cousteau began making underwater films. In 1950, he bought a boat named *Calypso,* which became his home and floating laboratory. For the next 40 years, through his films and television series, Cousteau brought what he called "the silent world" of the oceans and seas to living rooms everywhere.

A Protector of Life

Cousteau was long an outspoken defender of the environment. "When I saw all this beauty under the sea, I fell in love with it. And finally, when I realized to what extent the oceans were threatened, I decided to campaign as vigorously as I could against everything that threatened what I loved."

Jacques Cousteau died in 1997 at age 87. Before his death, he dedicated the *Calypso II,* a new research vessel, to the children of the world.

Write About It

▶ Ocean pollution and overfishing are subjects of intense debate. Think about these issues, and discuss them with your classmates. Then write an essay in which you try to convince readers of your point of view.

66

Answer to Write About It

Students should be able to find a great deal of information about ocean pollution and overfishing. Many articles have been published in various science magazines, and there is a large number of government and nonprofit Web sites that deal with these issues.

EYE ON THE ENVIRONMENT

Putting Freshwater Problems on Ice

Imagine how different your life would be if you couldn't get fresh water. What would you drink? How would you clean things? The Earth has enough fresh water to supply 100 billion liters to each person, yet water shortages affect millions of people every day. So what's the problem?

▲ *Icebergs such as this one might provide water in the future.*

The Ice-Water Planet

Three-quarters of Earth's fresh water is frozen in polar icecaps. Plenty of fresh water is there, but people can't use water that is frozen and thousands of kilometers away.

The ice sheet that covers Antarctica is thousands of meters thick and is almost twice the size of the United States. Hundreds of huge chunks break off its edges every year. These icebergs, which are made up entirely of frozen fresh water, float away into the sea and eventually melt. Water from 1 year's worth of these icebergs would be enough to supply all of southern California for more than a century. So why not use it?

Obvious but Not Easy

Transporting icebergs to areas that need fresh water is harder than it sounds. For one thing, many of the icebergs are huge. The largest ever recorded was about the size of Connecticut. Even small icebergs may be 2 km long and 1 km wide.

Researchers have considered many methods of transporting icebergs. Most of the ideas involve pushing or towing icebergs through the water. A few ideas involve attaching engines and propellers directly to the icebergs.

However, because icebergs are so large, it takes a long time to move them. And when an iceberg finally does get somewhere, a considerable amount of it has melted. To prevent melting, insulating materials could be wrapped around an iceberg.

A Worthy Investment

Lakes and ground water still provide the cheapest fresh water in most areas. However, if there is no lake, river, or well water available, icebergs may then be a reasonable option to consider. Even though transporting icebergs is difficult, it may still be worthwhile to try. Irrigating 100 km^2 of desert with water from icebergs might cost as much as $1 million, but purifying enough sea water to irrigate that amount of desert could cost over $1 billion.

People in arid regions have spent considerable time on iceberg research. So far, no one has set up a program for harvesting icebergs. But someday water from icebergs may flow from our household faucets.

An Icy Investigation

▶ Float an ice cube in a bowl of cold water, and record the time it takes the cube to melt. Then try to insulate other ice cubes with different materials, such as cloth, plastic wrap, and aluminum foil. Which material works best? How could this material be used on real icebergs?

EYE ON THE ENVIRONMENT

Putting Freshwater Problems on Ice

Background

Icebergs form in both the Northern and Southern Hemispheres, but they are typically very different in size and appearance. Icebergs in the Northern Hemisphere, such as those that break off from the ice sheet covering Greenland, tend to be spiky and irregularly shaped. Icebergs in the Southern Hemisphere are much larger and usually have a fairly flat top.

People first considered the possibility of using icebergs as a water source many years ago. Just after 1900, Antarctic icebergs were towed by steamships to Callao, Peru. In addition to using icebergs for fresh water, people have proposed using them for air conditioning, as a source of gourmet ice cubes, and even as giant aircraft carriers.

Answers to An Icy Investigation

Answers will vary. Students may find that some materials prevent the ice cubes from melting as quickly as they would if they were unprotected. Students should propose realistic ways of using their materials on icebergs.

Chapter Organizer

CHAPTER ORGANIZATION	TIME MINUTES	OBJECTIVES	LABS, INVESTIGATIONS, AND DEMONSTRATIONS
Chapter Opener pp. 68–69	45	National Standards: SAI 1, SPSP 3, 4, HNS 1, ES 1b	**Start-Up Activity,** When *Whirls* Collide, p. 69
Section 1 Currents	90	▶ Describe surface currents, and list the three factors that control them. ▶ Describe deep currents. ▶ Illustrate the factors involved in deep-current movement. ▶ Explain how currents affect climate. SAI 1, HNS 1, 3, ES 1j, 3d; Labs UCP 2, SAI 1	**Demonstration,** The Coriolis Effect, p. 72 **Interactive Explorations CD-ROM,** Latitude Attitude *A* ***Worksheet*** *is also available in the* ***Interactive Explorations Teacher's Edition.*** **Skill Builder,** Up from the Depths, p. 88 **Datasheets for LabBook,** Up from the Depths **Whiz-Bang Demonstrations,** Spin Cycle
Section 2 Waves	90	▶ Identify wave components, and explain how they relate to wave movement. ▶ Describe how ocean waves form and how they move. ▶ Classify types of waves. ▶ Analyze types of dangerous waves. UCP 3, SAI 1, SPSP 3, 4, ES 1b	**QuickLab,** Do the Wave, p. 82
Section 3 Tides	90	▶ Explain tides and their relationship with the Earth, the sun, and the moon. ▶ Classify different types of tides. ▶ Analyze the relationship between tides and coastal land. UCP 1, 2, HNS 1, 3, ES 3c; Labs UCP 2, SAI 1, ST 1, ES 3c	**Making Models,** Turning the Tides, p. 106 **Datasheets for LabBook,** Turning the Tides **Long-Term Projects & Research Ideas,** An Ocean Commotion

See page **T23** *for a complete correlation of this book with the*

NATIONAL SCIENCE EDUCATION STANDARDS.

TECHNOLOGY RESOURCES

Guided Reading Audio CD
English or Spanish, Chapter 3

One-Stop Planner CD-ROM with Test Generator

Interactive Explorations CD-ROM
CD 3, Exploration 4, Latitude Attitude

CNN. **Scientists in Action,** Mapping El Niño Erosion, Segment 20

Science Discovery Videodiscs
Science Sleuths: The Mystery Fog

Earth Science Videodisc
Tides: 23877–29491

Chapter 3 • The Movement of Ocean Water

CLASSROOM WORKSHEETS, TRANSPARENCIES, AND RESOURCES	SCIENCE INTEGRATION AND CONNECTIONS	REVIEW AND ASSESSMENT
Directed Reading Worksheet **Science Puzzlers, Twisters & Teasers**	**Careers:** Seismologist, p. 94	
Directed Reading Worksheet, Section 1 **Transparency 156,** Earth's Surface Currents **Transparency 156,** The Coriolis Effect **Math Skills for Science Worksheet,** What Is Scientific Notation? **Transparency 157,** How Deep Currents Form **Transparency 158,** Circulation of Deep and Surface Currents	**Real-World Connection,** p. 70 in ATE **Connect to Life Science,** p. 71 in ATE **Real-World Connection,** p. 72 in ATE **Physics Connection,** p. 73 **Math and More,** p. 74 in ATE **Connect to Life Science,** p. 75 in ATE **Cross-Disciplinary Focus,** p. 76 in ATE **Real-World Connection,** p. 77 in ATE	**Self-Check,** p. 71 **Section Review,** p. 73 **Homework,** pp. 73, 75 in ATE **Section Review,** p. 77 **Quiz,** p. 77 in ATE **Alternative Assessment,** p. 77 in ATE
Transparency 159, Wave Period **Directed Reading Worksheet,** Section 2 **Transparency 160,** Deep-Water Waves Become Shallow-Water Waves **Math Skills for Science Worksheet,** What Is a Fraction? **Transparency 282,** Wave Speed, Wavelength, and Frequency **Reinforcement Worksheet,** Waves to Your Pen Pal	**MathBreak,** Wave Speed, p. 79 **Math and More,** p. 79 in ATE **Real-World Connection,** p. 80 in ATE **Math and More,** p. 81 in ATE **Connect to Environmental Science,** p. 81 in ATE **Connect to Meteorology,** p. 82 in ATE **Apply,** p. 83	**Homework,** p. 81 in ATE **Section Review,** p. 83 **Quiz,** p. 83 in ATE **Alternative Assessment,** p. 83 in ATE
Directed Reading Worksheet, Section 3 **Critical Thinking Worksheet,** Tides of Trouble **Transparency 161,** Tidal Variations **Reinforcement Worksheet,** But What About the Tides?	**Connect to Environmental Science,** p. 87 in ATE **Health Watch:** Red Tides, p. 95	**Homework,** p. 86 in ATE **Section Review,** p. 87 **Quiz,** p. 87 in ATE **Alternative Assessment,** p. 87 in ATE

END-OF-CHAPTER REVIEW AND ASSESSMENT

Chapter Review in Study Guide
Vocabulary and Notes in Study Guide
Chapter Tests with Performance-Based Assessment, Chapter 3 Test
Chapter Tests with Performance-Based Assessment, Performance-Based Assessment 3
Concept Mapping Transparency 14

internetconnect

go.hrw.com

Holt, Rinehart and Winston On-line Resources

go.hrw.com

For worksheets and other teaching aids related to this chapter, visit the HRW Web site and type in the keyword: **HSTH2O**

sciLINKS NSTA

National Science Teachers Association

www.scilinks.org

Encourage students to use the *sci*LINKS numbers listed in the internet connect boxes to access information and resources on the **NSTA** Web site.

Chapter Resources & Worksheets

Visual Resources

TEACHING TRANSPARENCIES

#156

#157

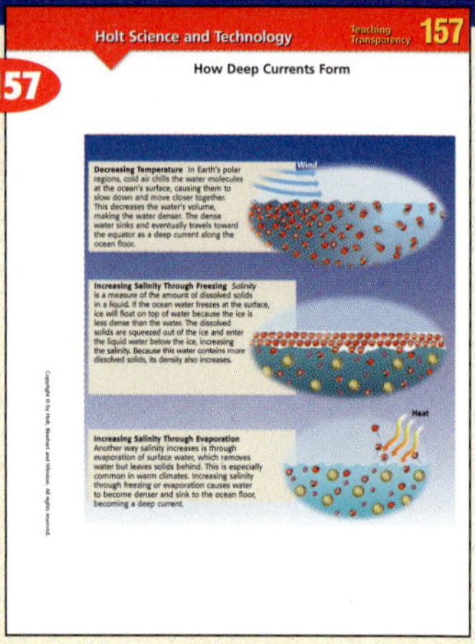

#158

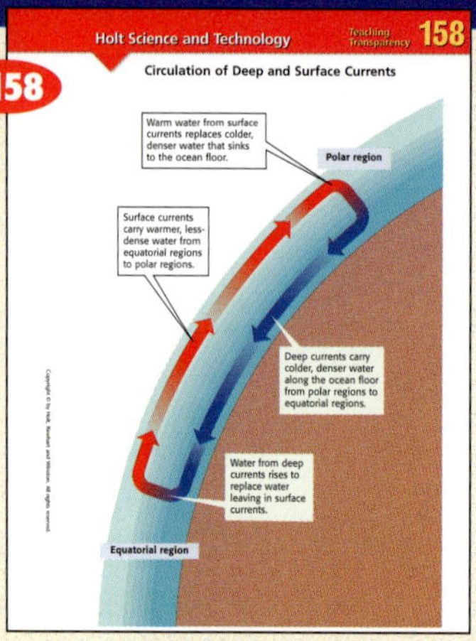

#159

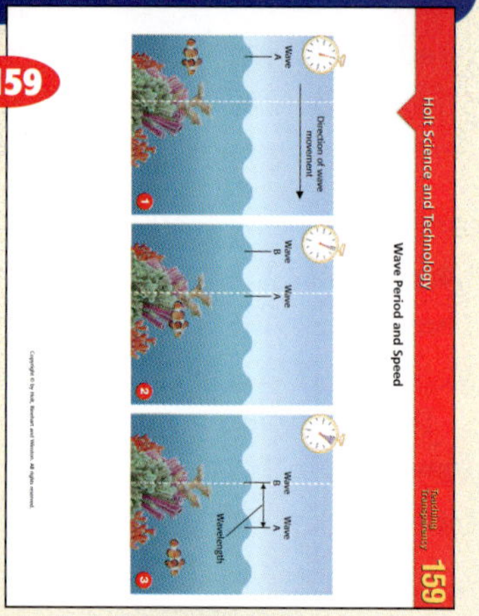

TEACHING TRANSPARENCIES

#160

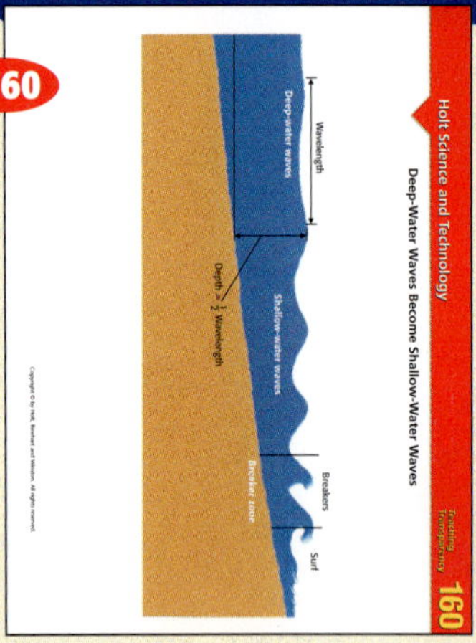

#161

#282

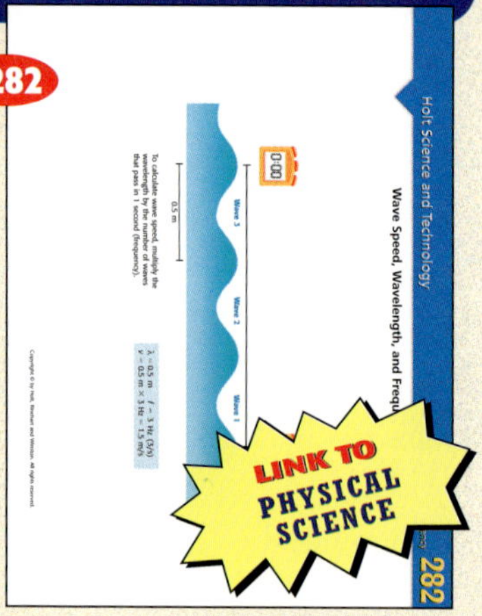

CONCEPT MAPPING TRANSPARENCY

#14

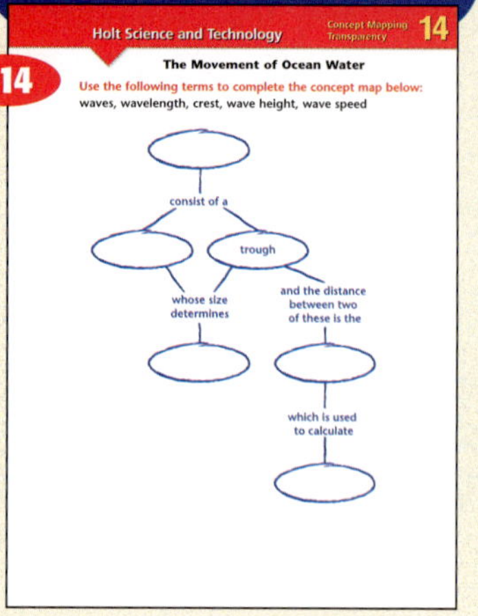

Meeting Individual Needs

DIRECTED READING

#3

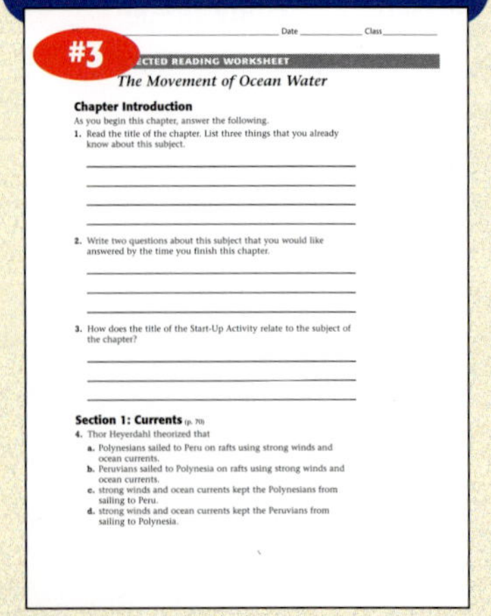

REINFORCEMENT & VOCABULARY REVIEW

#3

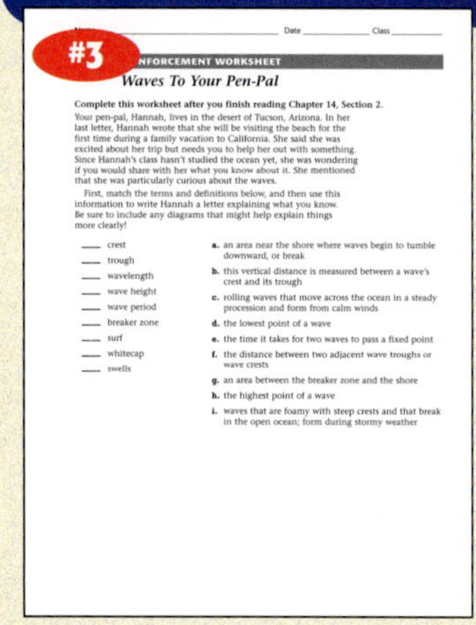

#3

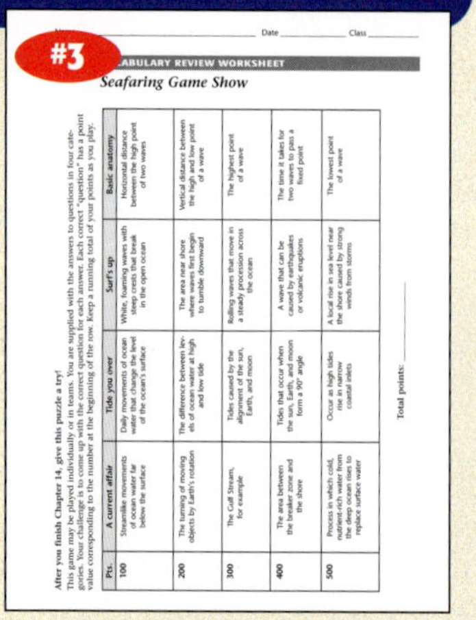

SCIENCE PUZZLERS, TWISTERS & TEASERS

#3

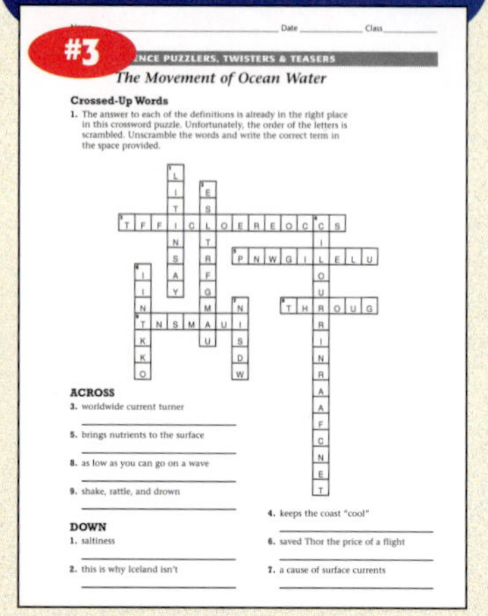

Chapter 3 • The Movement of Ocean Water

Review & Assessment

STUDY GUIDE

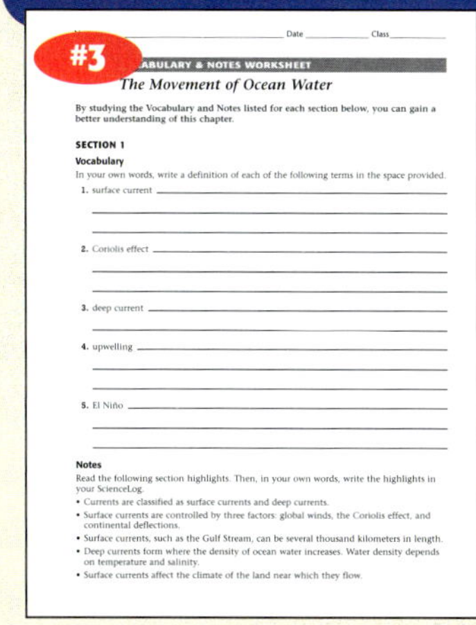
#3 Vocabulary & Notes Worksheet — *The Movement of Ocean Water*

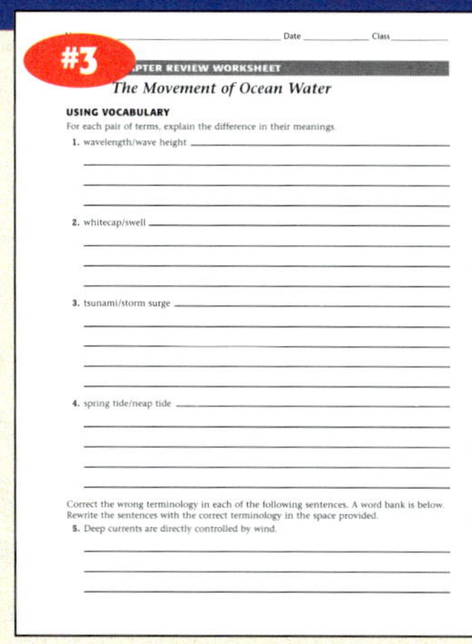
#3 Chapter Review Worksheet — *The Movement of Ocean Water*

CHAPTER TESTS WITH PERFORMANCE-BASED ASSESSMENT

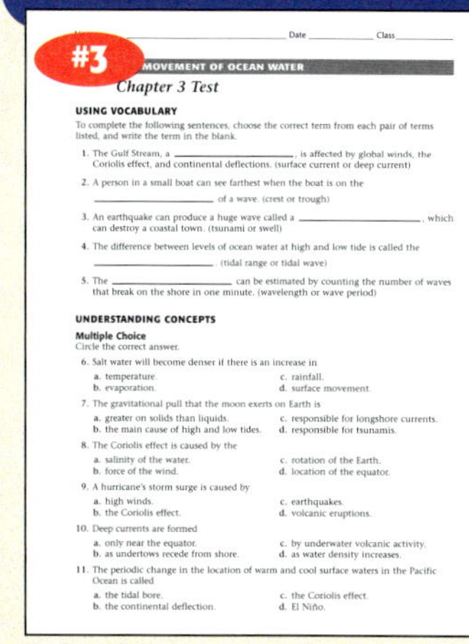
#3 The Movement of Ocean Water — *Chapter 3 Test*

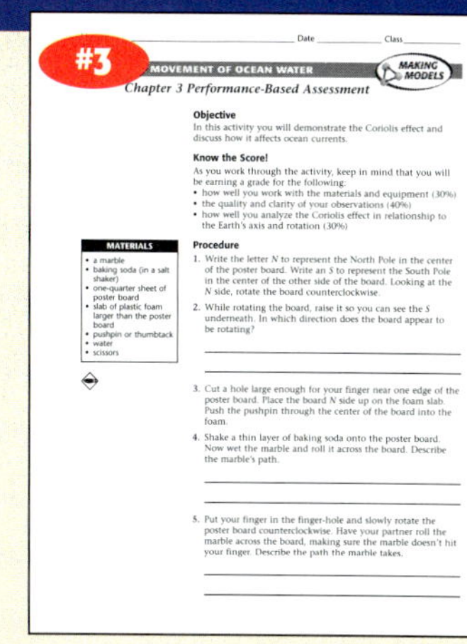
#3 The Movement of Ocean Water — *Chapter 3 Performance-Based Assessment*

Lab Worksheets

WHIZ-BANG DEMONSTRATIONS

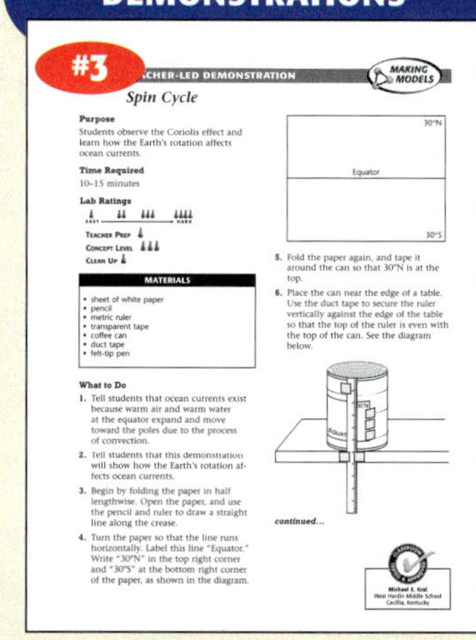
#3 Teacher-Led Demonstration — *Spin Cycle*

LONG-TERM PROJECTS & RESEARCH IDEAS

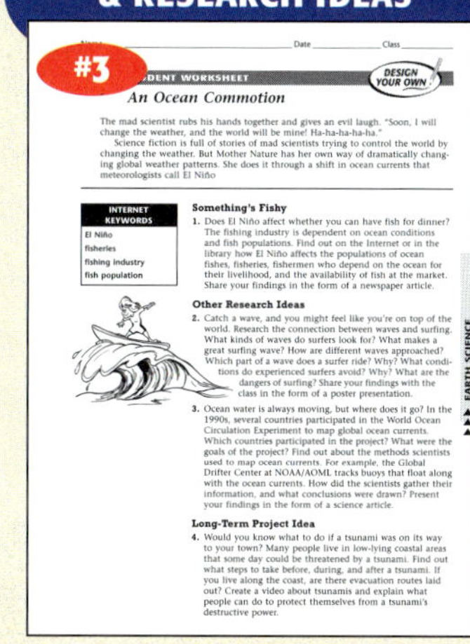
#3 Student Worksheet — *An Ocean Commotion*

DATASHEETS FOR LABBOOK

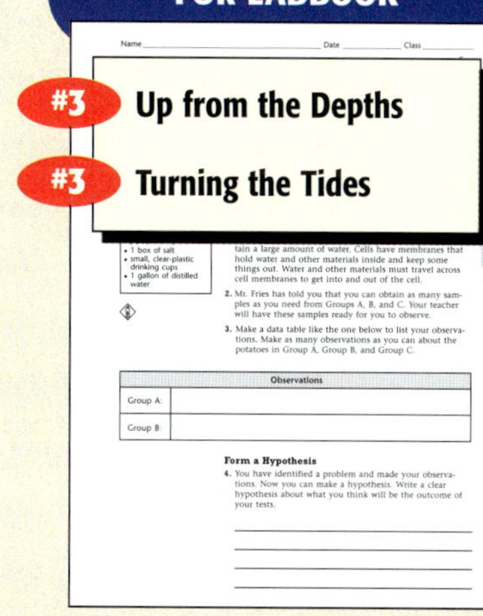
#3 Up from the Depths

#3 Turning the Tides

Applications & Extensions

CRITICAL THINKING & PROBLEM SOLVING

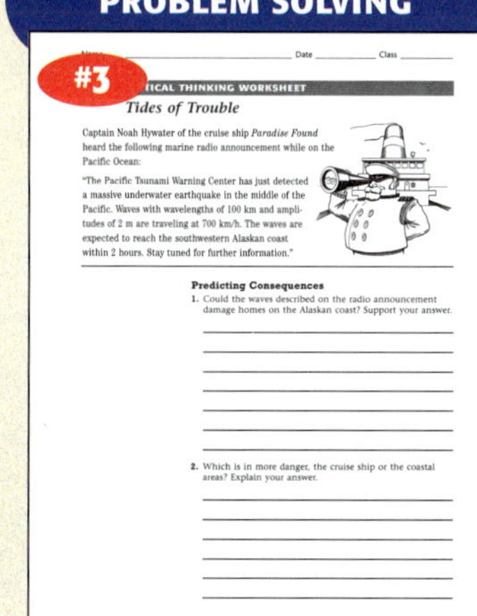
#3 Critical Thinking Worksheet — *Tides of Trouble*

SCIENTISTS IN ACTION

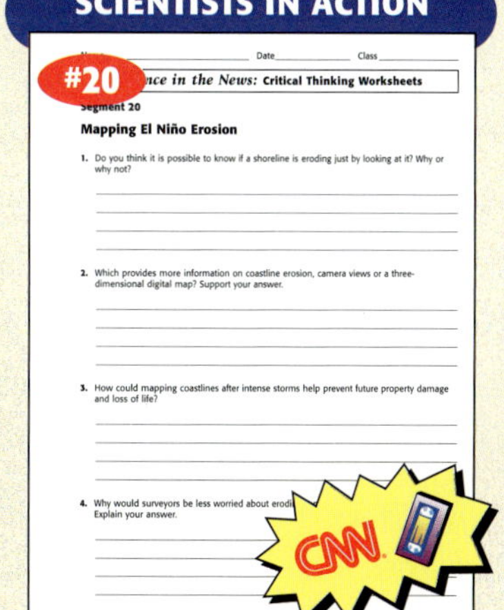
#20 Science in the News: Critical Thinking Worksheets — Segment 20 — Mapping El Niño Erosion

INTERACTIVE EXPLORATIONS

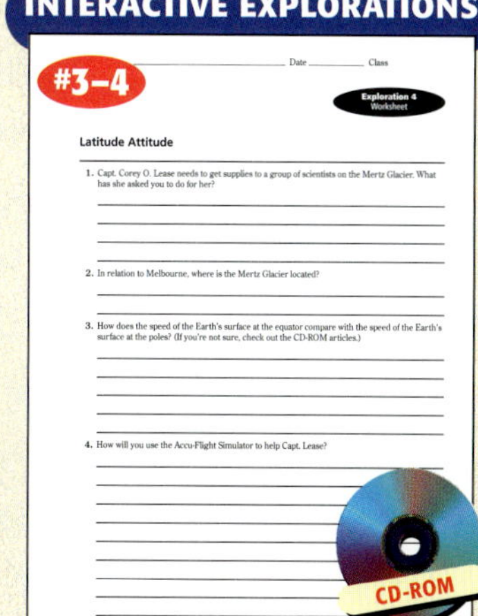
#3–4 Exploration 4 Worksheet — Latitude Attitude

Chapter Background

Section 1

Currents

Solar Radiation
One of the fundamental energy sources for all ocean currents is solar radiation. Uneven heating of the Earth by the sun creates differences in air pressure. These differences create wind, which drives surface currents. The sun's energy also creates temperature differences in ocean water, driving deep currents.

The *Ra II* Expedition
On May 17, 1970, Thor Heyerdahl's *Ra II* expedition attempted to demonstrate that mariners from ancient Egypt could have reached the New World. The eight-man expedition successfully reached its destination in the Barbados on July 12, 1970.

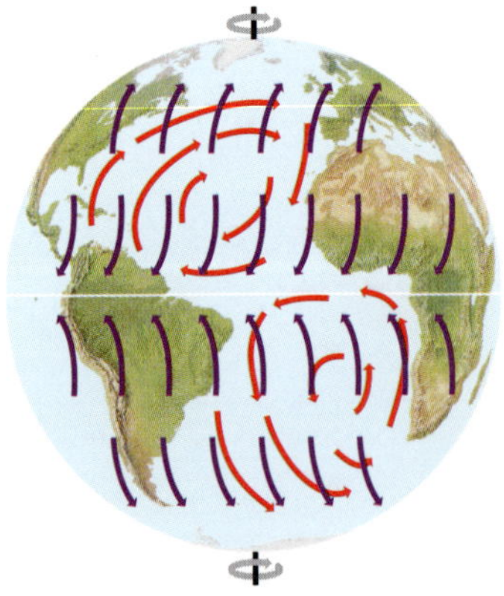

Gaspard Coriolis
The Coriolis effect was named after Gaspard Coriolis (1792–1843), a French engineer. Coriolis was the first to attribute the deflection of surface currents across the surface of Earth to a hypothetical force. It was not until later that the force was identified as Earth's rotation.

IS THAT A FACT!

- The strongest and largest ocean current is the Antarctic Circumpolar Current, which is estimated to flow at a rate of 125 million cubic meters per second.
- One of the fastest ocean currents is the Somali Current, in the western Indian Ocean, which flows at a speed of 14.5 km/h.
- The Weddell Sea, where the Antarctic Bottom Water is thought to originate, has the clearest water of any sea. Its clarity has been recorded to a depth of nearly 80 m. In other words, water collected from the upper 80 m of the Weddell Sea is as clear as you would find in a glass of distilled water.

Section 2

Waves

Swells
Swells are generated in the open ocean by wind and can travel thousands of kilometers to shore. These long-wavelength waves have periods of 10 to 30 seconds. When swells reach shallower water, their height increases, causing them to fall forward. When this occurs, the waves are called breakers.

Tsunamis

Tsunami is a Japanese word that means "harbor wave." Japan has experienced many devastating tsunamis throughout history. The subduction of tectonic plates generates the seismic energy necessary to cause tsunamis. Because the Japanese islands are on the edge of deep water and the coastline is rugged, with many small harbors, tsunamis have been particularly destructive. The deep water close to the Japanese islands keeps tsunami wave heights short until the waves are very close to shore; when the waves enter shallow water, they suddenly grow taller. The narrow shape of many Japanese harbors causes tsunamis to grow even taller.

- Tsunamis have the potential to be the most destructive of ocean waves. Their speed averages 500 km/h, and their period ranges from 5 to 60 minutes. Because the wave height of a tsunami is usually less than 2 m in the open ocean, they often pass unnoticed beneath ships.
- The earthquakes that produce destructive tsunamis are generally greater than 6.5 on the Richter scale. Most occur in the Pacific Ocean, where there is a high level of seismic activity near plate boundaries.

Section 3

Tides

Perigean Spring Tides

On March 7, 1993, a full moon appeared on the same day that the moon was closest to Earth in its orbit (or its perigee). This event produced extraordinarily high and low tides, called *perigean spring tides*. During perigean spring tides, high tides rise higher than normal onto beaches, while low tides expose parts of the ocean floor that are normally submerged.

Is That a Fact!

- The sun exerts only about half the tidal force on Earth that the moon does. This is because the distance between Earth and the sun is much greater than the distance between Earth and the moon.

- High tide that occurs on the side of Earth that faces the moon is called a *direct high tide*. High tide that occurs on the side of Earth opposite the moon is called an *indirect high tide*.
- The Great Lakes also experience tides. Although the tidal range is small compared with that of the oceans, the size and depth of the five lakes make small fluctuations in water level noticeable.

For background information about teaching strategies and issues, refer to the ***Professional Reference for Teachers.***

CHAPTER 3

The Movement of Ocean Water

Pre-Reading Questions

Students may not know the answers to these questions before reading the chapter, so accept any reasonable response.

Suggested Answers

1. Factors that control ocean currents include wind, the Earth's rotation, water density, water salinity, water temperature, and continental deflections.
2. Gravitational forces from the sun and moon and the motion of the Earth and the moon around each other cause the ocean tides.

CHAPTER 3

The Movement of Ocean Water

Sections

An Ocean Stream

The Gulf Stream current carries warm tropical water from the Caribbean Sea all the way to the North Atlantic Ocean. The climate in the British Isles, where the current ends, is controlled by the current's warm waters, which make the isles much warmer than other countries nearby. In this chapter, you will learn how currents like the Gulf Stream are formed. You also will learn about the other ways that ocean water moves and how these movements affect our lives.

Pre-Reading Questions

1. What factors control ocean currents?
2. What causes the ocean tides?

68

internet connect

HRW On-line Resources

go.hrw.com
For worksheets and other teaching aids, visit the HRW Web site and type in the keyword: **HSTH2O**

www.scilinks.com
Use the *sci*LINKS numbers at the end of each chapter for additional resources on the **NSTA** Web site.

www.si.edu/hrw
Visit the Smithsonian Institution Web site for related on-line resources.

www.cnnfyi.com
Visit the CNN Web site for current events coverage and classroom resources.

START-UP Activity

WHEN *WHIRLS* COLLIDE

Ocean currents in the Northern Hemisphere flow in a clockwise direction, while currents in the Southern Hemisphere flow in a counterclockwise direction. Sometimes, southern currents flow across the equator into the Northern Hemisphere and begin flowing clockwise. Do this activity to find out how currents flowing in opposite directions affect one another.

Procedure

1. Fill a large **tub** with **water** 5 cm deep.
2. Add **10 drops of red food coloring** to the water on one end of the tub.
3. Add **10 drops of blue food coloring** to the water at the other end of the tub.
4. Using a **pencil,** quickly stir the water at one end of the tub in a clockwise direction while your partner stirs the water at the other end in a counterclockwise direction. Stir both ends for about 5 seconds.
5. In your ScienceLog, draw what you see happening in the tub immediately after you stop stirring. (Both ends should still be swirling.)

Analysis

6. How did the blue water and the red water interact?
7. How does this activity relate to the ocean currents in the Northern and Southern Hemispheres?

START-UP Activity

WHEN *WHIRLS* COLLIDE

MATERIALS

For Each Pair of Students:
- large tub
- water
- red food coloring
- blue food coloring
- 2 pencils

Teacher's Notes

To avoid moving the tub after it is filled with water, put the tub in position before the water is added. Although food coloring is nontoxic and washes out of clothing, students should be careful not to spill the food coloring on their skin or clothes.

A cake pan similar in size to the tub can be used for this activity. Water depth should be about 5 cm. Advise students to observe the experiment closely—the desired result happens quickly and lasts only a few seconds.

Explore what happens with different types of tubs. Does the depth of the tub change the results?

Answers to START-UP Activity

6. When the blue water left the current it was circulating in, it crossed the middle of the tub, joined the red water, and began circling in the opposite direction. The red water followed the same pattern.
7. Answers will vary. Currents in the Northern Hemisphere circulate in the opposite direction as do currents in the Southern Hemisphere. When a current crosses the equator, it joins other currents and eventually circulates in the direction they are moving.

Section 1

Focus

Currents

This section discusses the causes and characteristics of surface and deep currents in the ocean. Students will learn about the different factors related to ocean currents and explore the ways currents and climate are related.

Bellringer

Read excerpts from Thor Heyerdahl's *Kon-Tiki* (1950) or *The Ra Expeditions* (1971) to students. Display a large map of the world that shows the different ocean currents. After showing students the origination and destination points for Heyerdahl's voyages, discuss which currents he would have used to reach his destinations. Point out that DNA testing later showed that Polynesians did not originate in Peru. Have students discuss how a scientific model might be successful, but not accurate. Sheltered English

1 Motivate

Real-World Connection

Have students discuss the characteristics of rivers. Ask students to compare rivers to surface currents. Lead students to the understanding that rivers and surface currents are very different but share some important similarities. For example, both are long, moving bodies of water. However, rivers flow due to the pull of gravity and are usually confined between banks. Surface currents are driven by the wind and the rotation of the Earth.

Section 1 Reading Warm-Up

Terms to Learn

surface current
Coriolis effect
deep current
upwelling
El Niño

What You'll Do

- Describe surface currents, and list the three factors that control them.
- Describe deep currents.
- Illustrate the factors involved in deep-current movement.
- Explain how currents affect climate.

Currents

Imagine that you are stranded on a desert island. You stuff a distress message into a bottle and throw it into the ocean, hoping it will find its way to someone who will send help. Is there any way to predict where your bottle may land?

One Way to Explore Currents

Earlier this century, a Norwegian explorer named Thor Heyerdahl tried to answer similar questions that involved human migration across the ocean. Heyerdahl theorized that the inhabitants of Polynesia originally sailed from Peru on rafts powered only by the wind and ocean currents. Unable to convince scientists of his theory, he decided to prove it. In 1947, Heyerdahl and a crew of five people set sail from Peru on a raft, as shown in **Figure 1.**

Figure 1 *The handcrafted* Kon-Tiki *was made mainly from materials that would have been available to ancient Peruvians.*

On the 97th day of their expedition, Heyerdahl and his crew landed on an island in Polynesia. Currents had carried the raft westward more than 6,000 km across the South Pacific. This supported Heyerdahl's theory that ocean currents carried the ancient Peruvians across the Pacific to Polynesia. Now let's take a closer look at currents. For example, what determines the direction in which a current moves? What forces create a current? Read on to learn the answers to these and other questions about currents.

70

Is That a Fact!

No metal was used in the construction of the *Kon-Tiki*. The wood raft was made of thick Peruvian balsa logs and featured a bamboo cabin set in the center, a large steering oar at the stern, and five centerboards. Two masts were used to support the rectangular sail. Although it crashed into a reef when it finally reached Polynesia, the *Kon-Tiki* was restored and is currently on display at a museum in Oslo, Norway.

Surface Currents

Streamlike movements of water that occur at or near the surface of the ocean are called **surface currents.** Some surface currents are several thousand kilometers in length, traveling across entire oceans. The Gulf Stream, which is one of the longest surface currents, transports 25 times more water than all the rivers in the world. Surface currents are controlled by three factors: global winds, the Coriolis effect, and continental deflections. These three factors keep surface currents flowing in distinct patterns around the Earth.

Global Winds Have you ever blown gently on a cup of hot chocolate? You may have noticed ripples moving across the surface. These ripples are caused by a tiny surface current created by your breath. In much the same way, winds blowing across the Earth's surface create surface currents in the ocean. Surface currents can reach depths of several hundred meters and lengths of several thousand kilometers.

Different winds cause currents to flow in different directions. Near the equator, the winds blow ocean water east to west, but closer to the poles, ocean water is blown west to east, as shown in **Figure 2.** Merchant ships often use these currents to travel more quickly back and forth across the oceans.

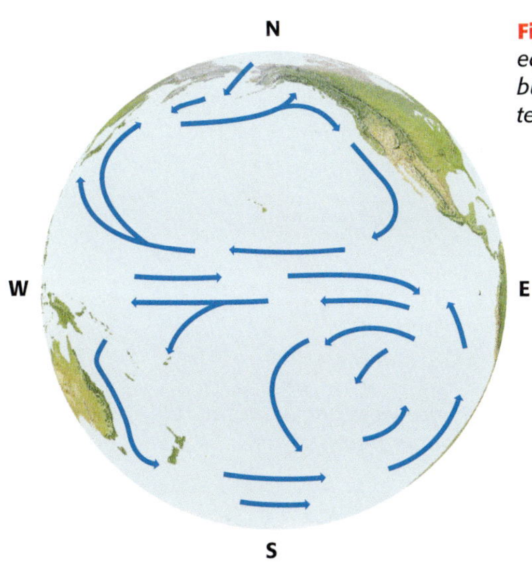

Figure 2 *Surface currents near the equator generally flow from east to west, but surface currents closer to the poles tend to flow from west to east.*

Self-Check

Take another look at Figure 2. As Heyerdahl made his journey in 1947, from what direction would he have noticed the wind blowing? *(See page 136 to check your answer.)*

71

Answer to Self-Check

It may help students to look at a globe when answering the Self-Check. The Polynesian Islands are very small islands in the South Pacific, and students may not know where Peru is located in South America. Because he was traveling from Peru to a Polynesian island to the west, Heyerdahl would have noticed the wind blowing from the east.

2 Teach

Reading Strategy

Prediction Guide Before students read the passage describing the three causes of surface currents, ask them what they think might cause currents on the surface of oceans. Students will discover the answers as they continue reading Section 1.

Connect to Life Science

Writing — Ocean currents help animals and plants colonize new islands. The Galápagos Islands are 1,100 km west of South America. Of the reptiles on the islands, 20 of the 22 species are found nowhere else in the world. Scientists speculate that the ancestors of these reptiles found their way from South America to the islands on natural rafts carried by the Humbolt and South Equatorial Currents. (The same currents that carried Thor Heyerdahl to Polynesia.) Reptiles such as lizards, snakes, and iguanas sometimes ride these rafts until they wash up on the shores of distant islands. Have interested students find out more about the unique plants and animals of the Galápagos Islands.

Directed Reading Worksheet Section 1

internetconnect

SciLinks NSTA

TOPIC: Ocean Currents
GO TO: www.scilinks.org
***sci*LINKS NUMBER:** HSTE330

2 Teach, continued

Reteaching

The Coriolis Effect For students having difficulty understanding the Coriolis effect, use a turntable to demonstrate the concept.

1. Cover the surface of the turntable with a circle of paper cut to fit.
2. Spin the turntable platter and explain that it represents the rotating Earth.
3. Instruct a student to attempt to draw a straight line from the center of the turntable to the edge. A curved line will be formed. The curved line represents the curved path of surface currents due to Earth's rotation.

Sheltered English

Real-World Connection

Surface currents move in large, slow circles called *gyres.* In 1990, a Korean ship carrying a load of athletic shoes was bound for the United States. During a storm, 80,000 shoes were washed overboard. Six months later, the shoes began to wash up on North American shores from Oregon to British Columbia. This accident turned out to be a big bonus to oceanographers. They created a computer model to predict where the gyres would carry the shoes next. True to the model, the shoes began to wash up in Hawaii three years later. With the help of dedicated beachcombers, this accident has enabled oceanographers to create a detailed map of surface currents in the Pacific Ocean. Have students look at a map of global ocean currents and identify the currents that carried the shoes. Students may also enjoy visiting Web sites that track the movements of other cargo spills.

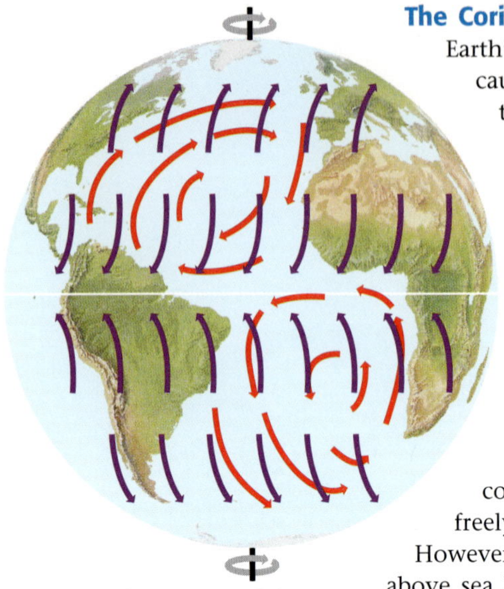

Figure 3 *The rotation of the Earth causes ocean currents (red arrows) and global winds (purple arrows) to move in opposite directions on either side of the equator.*

The Coriolis Effect Have you ever thought about how the Earth's rotation affects its surface? The Earth's rotation causes surface currents to move in curved paths rather than in straight lines. The curving of moving objects from a straight path due to the Earth's rotation is called the **Coriolis effect.** To understand the Coriolis effect, imagine trying to roll a ball straight across a turning merry-go-round. Because the merry-go-round is spinning, the path of the ball will curve before it reaches the other side. **Figure 3** shows that ocean currents in the Northern Hemisphere turn clockwise, while ocean currents in the Southern Hemisphere turn counterclockwise.

Continental Deflections If the Earth's surface were covered only with water, surface currents would travel freely across the globe in a very uniform pattern. However, we know that this is not the case—continents rise above sea level over roughly one-third of the Earth's surface. When surface currents meet continents, they *deflect,* or change direction. Notice in **Figure 4** how the Brazil Current deflects southward as it meets the east coast of South America.

Activity

Some people think the Coriolis effect can be seen in sinks; that is, water draining from sinks turns clockwise in the Northern Hemisphere and counterclockwise in the Southern Hemisphere. Is this true? Research this question at the library, on the Internet, and in your sinks and tubs at home.

TRY at HOME

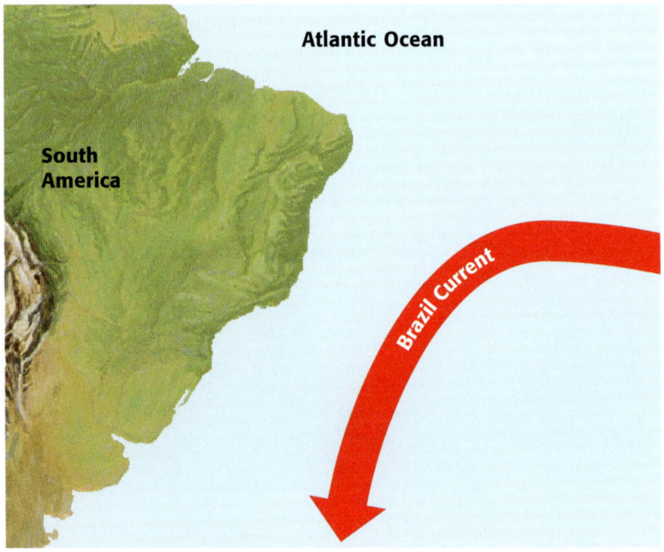

Figure 4 *If South America were not in the way, the Brazil Current would probably flow farther west.*

72

Misconception Alert

It is a common misconception that the direction in which water drains (clockwise versus counterclockwise) is determined by the Coriolis effect. Allow students to discover this themselves by directing them to do the Activity on this page.

Is That a Fact!

The Earth's rotation affects ocean currents directly by causing currents to circle in opposite directions on either side of the equator. Earth's rotation also affects the wind patterns, which in turn drive surface currents.

Taking Temperatures All three factors—global winds, the Coriolis effect, and continental deflections—work together to form a pattern of surface currents on Earth. But currents are also affected by the temperature of the water in which they arise. Warm-water currents begin near the equator and carry warm water to other parts of the ocean. Cold-water currents begin closer to the poles and carry cool water to other parts of the ocean. As you can see on the map in **Figure 5,** all the oceans are connected, and both warm-water and cold-water currents travel from one ocean to another.

Physics CONNECTION

While winds are responsible for ocean currents, the sun is the initial energy source of the currents. Because the sun heats the Earth more in some places than in others, convection currents are formed, which cause winds to blow.

Figure 5 *This map shows Earth's surface currents. Warm-water currents are shown as red arrows, and cold-water currents are shown as blue arrows.*

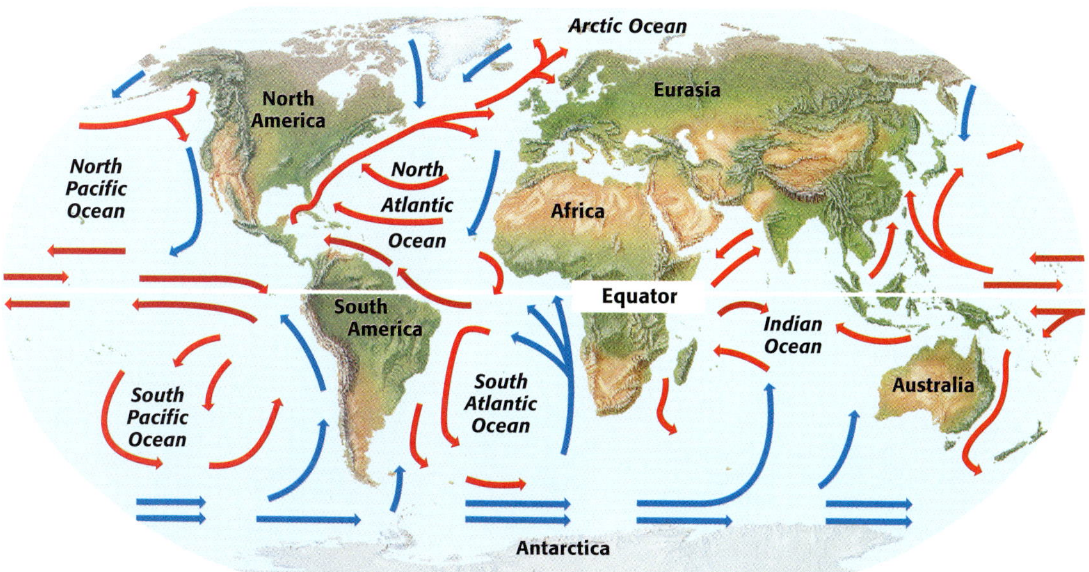

SECTION REVIEW

1. List the three factors that control surface currents.
2. Explain how the Earth's rotation affects the patterns of surface currents.
3. **Inferring Conclusions** If there were no land on Earth's surface, what would the pattern of surface currents look like? Explain.

internetconnect

SCLINKS. NSTA

TOPIC: Ocean Currents
GO TO: www.scilinks.org
***sci*LINKS NUMBER:** HSTE330

73

MEETING INDIVIDUAL NEEDS

Learners Having Difficulty
Refer to **Figure 5** to point out that the currents in the Southern Hemisphere turn counterclockwise and the currents in the Northern Hemisphere turn clockwise. Using a globe, show the direction of ocean currents in various areas around the world. Note for students that a current in the South Atlantic Ocean crosses the equator and joins the clockwise-turning currents. Discuss how this process was modeled in the Investigate activity at the beginning of this chapter. Sheltered English

Homework

Research Have students choose four coastal cities that are separated by oceans. Ask students to describe in their ScienceLog the currents they would use to sail from one city to the next. Students should use **Figure 5** as a reference for determining which currents they would use. Students can use an atlas or another reference tool to find the names of their currents. They can write their entries in the form of a ship's log.

Teaching Transparency 156
"Earth's Surface Currents"
"The Coriolis Effect"

Answers to Section Review

1. The three factors that control surface currents are global winds, the Coriolis effect, and continental deflections.
2. The Earth's rotation causes the Coriolis effect. The Coriolis effect is seen in the pattern of the ocean's surface currents—surface currents in the Northern Hemisphere turn clockwise, and surface currents in the Southern Hemisphere turn counterclockwise.
3. Answers will vary. Sample answer: If there were no land on Earth's surface, surface currents would not deflect sharply, as they do when they approach continents.

2 Teach, *continued*

Reading Strategy

Prediction Guide Before students read the section on deep currents, ask them to respond to the following questions:

Do global winds directly cause deep currents? (no)

Can ocean currents be caused by differences in water temperature? (yes)

Can ocean currents be caused by differences in salinity? (yes)

MATH and MORE

Review scientific notation with students. Tell them there are approximately 1.7×10^{19} water molecules in one drop of sea water and 1×10^{44} water molecules in the Mediterranean Sea. Have students help you write these two numbers to show how large they are. For reference, tell students that there are 17 times as many molecules in a drop of water than there are insects on Earth.

Math Skills Worksheet "What Is Scientific Notation?"

Teaching Transparency 157 "How Deep Currents Form"

Deep Currents

Deep currents are streamlike movements of ocean water far below the surface. Unlike surface currents, deep currents are not directly controlled by wind or the Coriolis effect. Instead, they form in parts of the ocean where water density increases. *Density* is the ratio of the mass of a substance to its volume. Two main factors—temperature and salinity—combine to affect the density of ocean water, as shown below. As you can see, both decreasing the temperature of ocean water and increasing the water's salinity increase the water's density.

How Deep Currents Form

Decreasing Temperature In Earth's polar regions, cold air chills the water molecules at the ocean's surface, causing them to slow down and move closer together. This decreases the water's volume, making the water denser. The dense water sinks and eventually travels toward the equator as a deep current along the ocean floor.

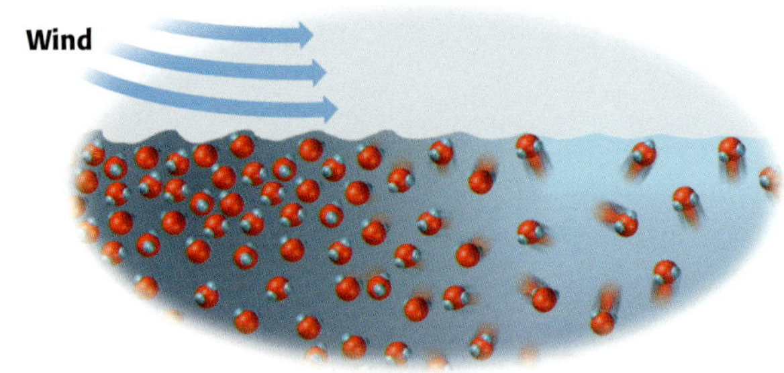

Increasing Salinity Through Freezing *Salinity* is a measure of the amount of dissolved solids in a liquid. If the ocean water freezes at the surface, ice will float on top of water because ice is less dense than liquid water. The dissolved solids are squeezed out of the ice and enter the liquid water below the ice, increasing the salinity. Because this water contains more dissolved solids, its density also increases.

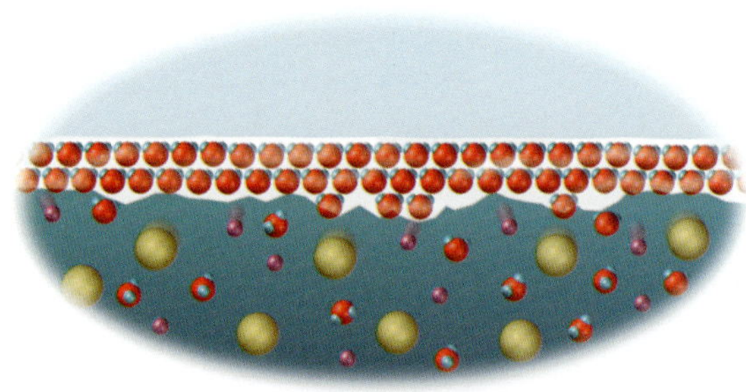

Increasing Salinity Through Evaporation Another way salinity increases is through evaporation of surface water, which removes water but leaves solids behind. This is especially common in warm climates. Increasing salinity through freezing or evaporation causes water to become denser and sink to the ocean floor, becoming a deep current.

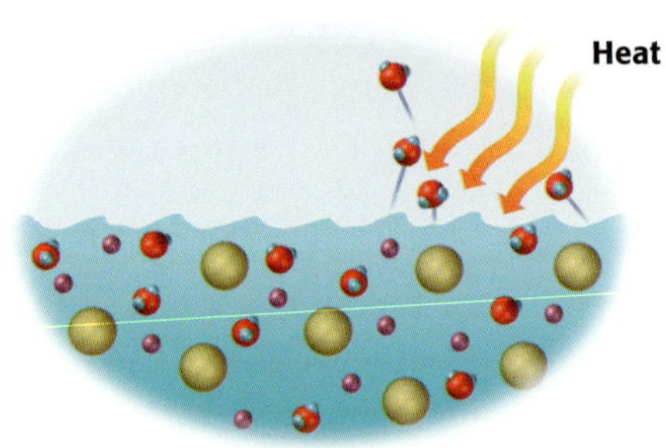

MISCONCEPTION ALERT

Remind students that molecules aren't made of little colored balls—the balls in the illustration on this page are models that represent molecules. The red and blue balls that are attached to one another represent water molecules, and the other balls represent dissolved solids.

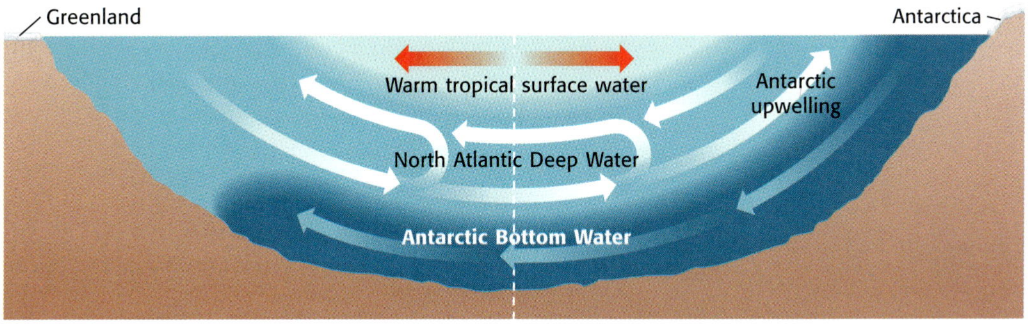

Figure 6 *Less-dense water always flows on top of denser water, as shown in this cross section.*

Movement of Deep Currents The movement of deep currents as they travel along the ocean floor is very complex. Differences in temperature and salinity, and therefore in density, cause variations in deep currents. For example, the deepest current, the Antarctic Bottom Water, is denser than the North Atlantic Deep Water. Both currents spread out across the ocean floor as they flow toward the same equatorial region. But when the currents meet, the North Atlantic Deep Water actually flows on top of the denser Antarctic Bottom Water, as shown in **Figure 6.** The Antarctic Bottom Water is so dense that it moves incredibly slowly—it takes 750 years for water in this current to make it from Antarctica's coastal waters to the equator!

Currents Trading Places Now that you understand how deep currents form and how they move along the ocean floor, you can learn how they trade places with surface currents. To see how this works, study **Figure 7.**

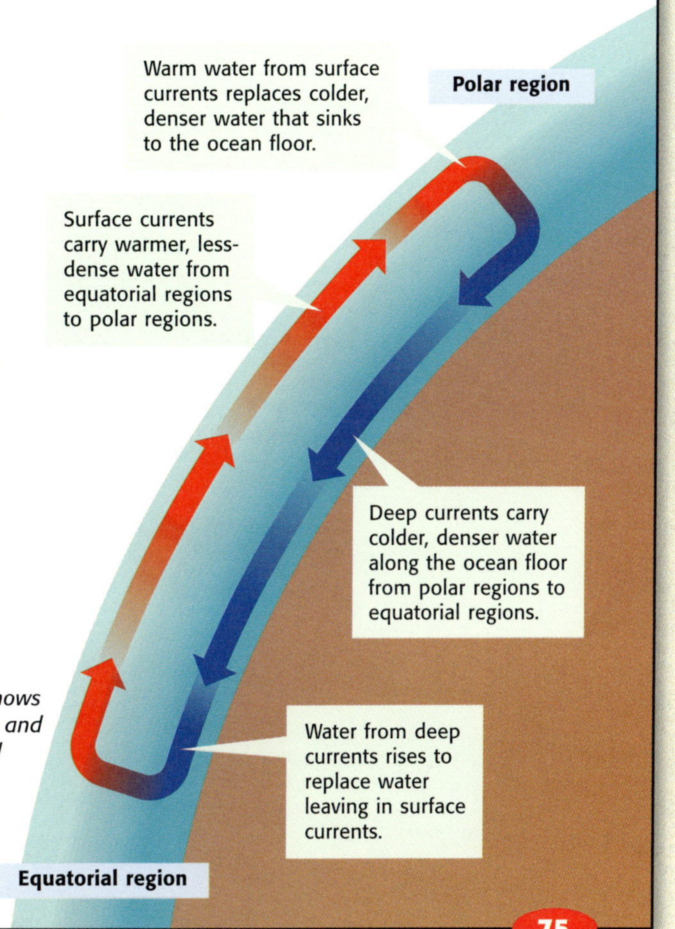

Figure 7 *This cross section shows the movement of warm water and cold water between polar and equatorial regions.*

Using the Figure

Figure 6 and **Figure 7** may appear to contradict one another. Explain to students that **Figure 6** is a detailed diagram showing the interaction of specific currents, while **Figure 7** is a generalized description of a cycle involving deep currents and surface currents. Make sure students realize that colder water becomes warmer near the equator and that warmer water becomes colder near the poles.

CONNECT TO LIFE SCIENCE

Along some coastlines surface currents diverge, carrying water away from the shore. These flow patterns cause upwelling, the rise of colder deep water that contains many nutrients required by phytoplankton. Where the nutrients reach the surface waters, phytoplankton growth flourishes. Phytoplankton are food for many species of commercial fish. Upwelling is common near the equator and in many coastal zones, usually on the western margins of continents. Populations in these coastal areas depend on the fishing industry as a major source of income and food. Encourage interested students to find out about different species of commercial fish harvested off the western coasts of North and South America and how El Niño affects upwelling.

Homework

Graphing Have students select two pairs of cities on opposite sides of a continent. All the cities should be at approximately the same latitude. One city from each pair should be located on the coast, and the other for each pair should be less than 300 km inland. For example, they could choose San Francisco and Fresno, California, and Norfolk and Roanoke, Virginia. Have students find the average high and low temperatures for each city. Next, have students find the average ocean temperature at the coastal cities. Have students plot the temperatures on a bar graph. Ask students to explain how ocean temperature affects the climate of the coastal cities.

Teaching Transparency 158
"Circulation of Deep and Surface Currents"

Surface Currents and Climate

Surface currents greatly affect the climate in many parts of the world. Some surface currents warm or cool coastal areas year-round. Other surface currents sometimes change their circulation pattern. This causes changes in the atmosphere that disrupt the climate in many parts of the world.

Currents That Stabilize Climate Although surface currents are generally much warmer than deep currents, their temperatures do vary. Surface currents are classified as warm-water currents or cold-water currents. Look back at Figure 5 to see where each type is located. Because they are warm or cold, surface currents affect the climate of the land near the area where they flow. For example, warm-water currents create warmer climates in coastal areas that would otherwise be much cooler. Likewise, cold-water currents create cooler climates in coastal areas that would otherwise be much warmer. **Figure 8** shows how a warm-water current and a cold-water current affect coastal climates.

Figure 8 *Warm-water currents, such as the Gulf Stream (top), and cold-water currents, such as the California Current (bottom), can affect the climate of coastal regions.*

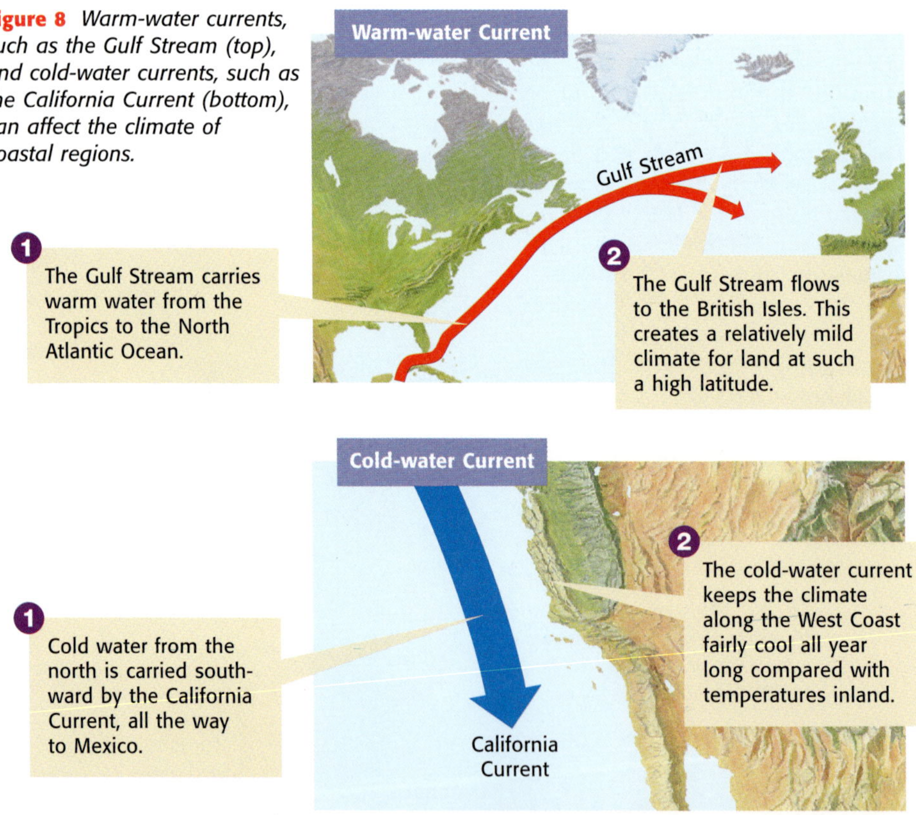

76

3 Extend

Cross-Disciplinary Focus

History Early in American history, Benjamin Franklin noticed that mail ships took much longer to travel from England to America than from America to England. He then discovered that ships from England were sailing against the Gulf Stream. Franklin revolutionized ocean navigation by designing charts that helped sailors avoid sailing against major surface currents.

Brain Food

Hurricanes are large tropical storms that originate over large bodies of warm water, such as the Atlantic Ocean or the Caribbean Sea. Hurricanes are created when warm water builds storm clouds, resulting in a massive low-pressure cell. In North America, hurricanes generally make landfall on the East Coast of the United States or the Gulf of Mexico, and they have reached as far north as Maine. On the West Coast, a hurricane has never made landfall in California. Ask students:

Why does the California Current, shown in **Figure 8,** protect California from hurricanes? (Because it is a cold-water current; the water off the coast of California is not warm enough to generate storm clouds or a low-pressure cell.)

Misconception Alert

El Niño is not a storm or any kind of weather pattern. It is a periodic change in the location of warm and cool surface waters in the Pacific Ocean. However, El Niño does ultimately cause local weather conditions to differ from typical conditions.

internetconnect

SciLINKS NSTA
TOPIC: El Niño
GO TO: www.scilinks.org
***sci*LINKS NUMBER:** HSTE335

Current Variations—El Niño The surface currents in the tropical region of the Pacific Ocean usually travel with the trade winds from east to west. This builds up warm water in the western Pacific and causes upwelling in the eastern Pacific. **Upwelling** is a process in which cold, nutrient-rich water from the deep ocean rises to the surface and replaces warm surface water. The warm water is blown out to sea by prevailing winds. But every 2 to 12 years, the South Pacific trade winds move less warm water to the western Pacific. As a result, surface water temperatures along the coast of South America rise. Gradually, this warming spreads westward. This periodic change in the location of warm and cool surface waters in the Pacific Ocean is called **El Niño.** El Niño not only affects surface waters but also changes the interaction between the ocean and the atmosphere, resulting in changes in global weather patterns.

BRAIN FOOD

El Niño means "The Child" in Spanish. The term originally referred to a warm current that arrived each year during the Christmas season off the coast of Ecuador and Peru.

Effects of El Niño El Niño alters weather patterns enough to cause disasters, such as flash floods and mudslides in areas of the world that normally receive little rain. **Figure 9** shows homes destroyed by a mudslide in Southern California. While some regions flood, regions that usually get a lot of rain may experience droughts, which can lead to crop failures.

Figure 9 *This damage in Southern California was the result of excessive rain caused by El Niño in 1997.*

SECTION REVIEW

1. How do temperature and salinity relate to deep-current movement?
2. Why is the climate in Scotland relatively mild even though the country is located at a high latitude?
3. **Applying Concepts** Many marine organisms depend on upwelling to bring nutrients to the surface. How might an El Niño affect Peruvians' way of life?

internetconnect

SCI LINKS. NSTA

TOPIC: El Niño
GO TO: www.scilinks.org
***sci*LINKS NUMBER:** HSTE335

Answers to Section Review

1. Decreasing the water's temperature or increasing its salinity will increase its density. As water becomes more dense, it sinks and becomes a deep current that moves along the ocean floor.
2. Because the Gulf Stream brings warm water from the tropics to Scotland's shore, the thermal energy retained in the water helps warm the air in Scotland.
3. Sample answer: El Niño prevents upwelling from occurring along the coast of Peru. Without upwelling, nutrients from deep water do not rise to the surface. Marine organisms that depend on the nutrients for food may die. Peruvians who depend on these organisms may suffer. For example, if fish die, the fishing industry will suffer.

REAL-WORLD CONNECTION

El Niño has far-reaching effects on many countries. Many areas have been devastated as a result of El Niño–related floods, storms, and droughts. Have students search for news stories describing some of the effects of El Niño in recent years. Encourage students to contrast these reports with the positive effects of El Niño, such as extended growing seasons.

4 Close

Quiz

Ask students to give two characteristics and one example of each type of current.

1. Sample answer: Surface currents occur at the surface of the ocean and are influenced by the Coriolis effect; the Gulf Stream is an example.
2. Sample answer: Deep currents occur deep in the ocean and are influenced by differences in water density; the Antarctic Bottom Water is an example.

ALTERNATIVE ASSESSMENT

PORTFOLIO Ask students to imagine they are planning a voyage around the world. They can choose any route they wish, but they must sail with the currents. Have them map out their selected route, showing the names of the currents, their point of origin, and point of destination.

Interactive Explorations, CD-ROM "Latitude Attitude"

SECTION 2

Focus

Waves

This section describes the characteristics of waves. Students will explore wave formation and movement. They will also learn how to identify different types of waves and measure different wave features. Finally, the section discusses dangerous movements of ocean water.

Bellringer

Illustrate the following scenario on the board or overhead projector:

You are floating in the ocean 1 km from shore, which is north of you. There is a surface current flowing east. Are you more likely to travel north with the waves toward the shore or east with the surface current? (east, because wave energy travels through the water but the water doesn't travel with the waves)

Tell students that Section 2 will help explain this answer.

1 Motivate

ACTIVITY

Arrange chairs in a long row, and have students sit in the chairs. Then have them stand and sit in succession to form a "human wave." Ask them to discuss and then demonstrate how the shape and motion of the wave could be changed. (Examples include standing and sitting more quickly to decrease the wave period or stretching their arms over their heads as they stand and bringing them down to their sides as they sit to increase the wave height.) Sheltered English

SECTION 2 READING WARM-UP

Terms to Learn

crest	surf
trough	whitecap
wavelength	swells
wave height	tsunami
wave period	storm surge
breaker	

What You'll Do

- Identify wave components, and explain how they relate to wave movement.
- Describe how ocean waves form and how they move.
- Classify types of waves.
- Analyze types of dangerous waves.

Waves

We all know what ocean waves look like. Even if you've never been to the seashore, you've most likely seen waves on television. But what are ocean waves? How do they form and move? Are all waves the same? And what do they do besides drop shells and sand dollars on the beach? Let us examine ocean waves so that we can answer these questions.

Anatomy of a Wave

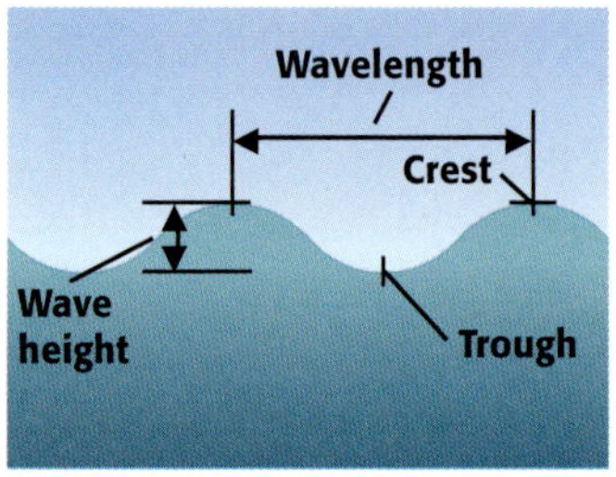

Waves are made up of two main components—crests and troughs. A **crest** is the highest point of a wave, and a **trough** is the lowest point. Imagine a thrilling roller coaster designed with many rises and dips. The top of a rise on a roller-coaster track is similar to the crest of a wave, and the bottom of a dip in the track resembles the trough of a wave. The distance between two adjacent wave crests or wave troughs is a **wavelength.** The vertical distance between a wave's crest and its trough is a **wave height.**

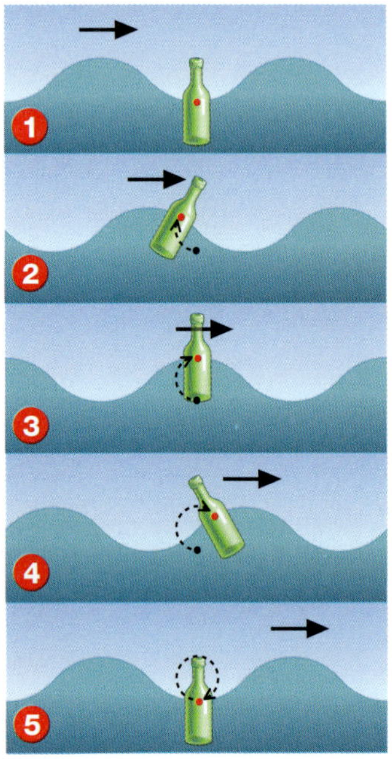

Figure 10 *Like the bottle in this figure, water remains in the same place as waves travel through it.*

Wave Formation and Movement

If you have watched ocean waves before, you may have noticed that water appears to move across the ocean's surface. However, this movement is only an illusion. Most waves form as wind blows across the water's surface, transferring energy to the water. As the energy moves through the water, so do the waves. But the water itself stays behind, rising and falling in circular movements. Notice in **Figure 10** that the floating bottle remains in the same spot as the waves travel from left to right. The circle of moving water that the bottle moves with has a diameter that is equal to the height of the waves that created it. Underneath this circle are smaller circles of moving water. The diameters of these circles get smaller with depth because wave energy decreases with depth. Wave energy only reaches to a certain depth. Below that depth, the water is not affected by wave energy.

78

Draw this diagram to show students how the circular motion of water in waves decreases in size as depth increases.

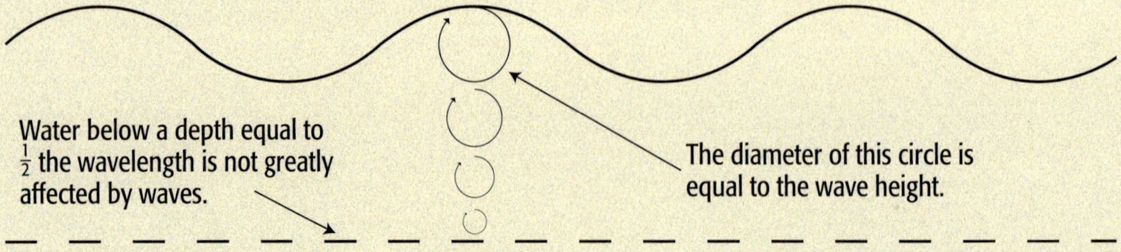

Specifics of Wave Movement

Waves not only come in different sizes but also travel at different speeds. To calculate wave speed, scientists must know the wavelength and the wave period. **Wave period** is the time between the passage of two wave crests (or troughs) at a fixed point, as shown in **Figure 11.** Dividing wavelength by wave period gives you wave speed, as shown below.

$$\frac{\text{wavelength (m)}}{\text{wave period (s)}} = \text{wave speed (m/s)}$$

For any given wavelength, an increase in the wave period will decrease the wave speed, and a decrease in the wave period will increase the wave speed.

MATH BREAK

Wave Speed

Imagine you are in a rowboat on the open ocean. You count 2 waves traveling right under your boat in 10 seconds. You estimate the wavelength to be 3 m. What is the wave speed?

Figure 11 *The illustration below shows how the wave period is determined.*

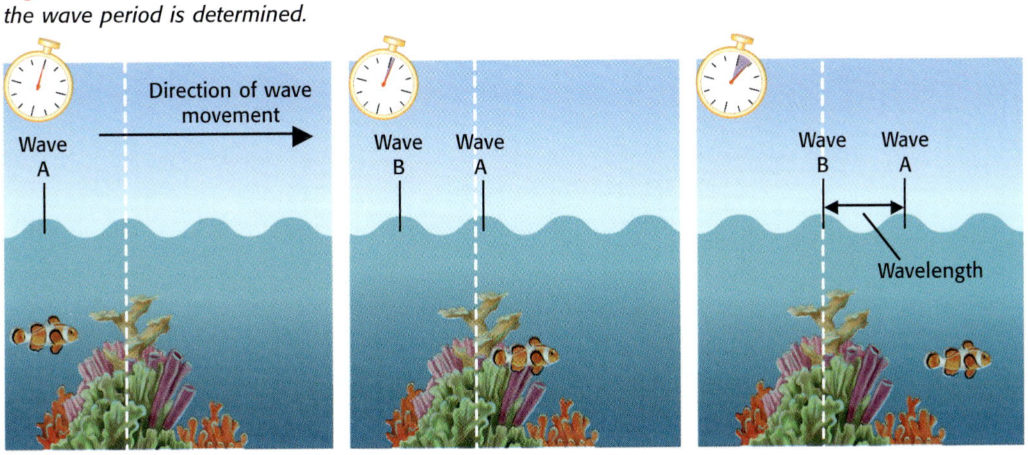

1 *Notice that the waves are moving from left to right.*

2 *The clock begins running as Wave A passes the reef's peak.*

3 *The clock stops as Wave B passes the reef's peak. The time shown on the clock (5 seconds) represents the wave period.*

Types of Waves

As you learned earlier in this section, wind forms most ocean waves. However, waves can form by other mechanisms. Underwater earthquakes and landslides as well as impacts by cosmic bodies can form different types of waves. The sizes of the different types of waves can vary, but most move the same way. Depending on their size and the angle at which they hit the shore, waves can generate a variety of near-shore events, some of which can be dangerous to humans.

internet connect

SCILINKS NSTA

TOPIC: Ocean Waves
GO TO: www.scilinks.org
***sci*LINKS NUMBER:** HSTE340

TOPIC: Tsunamis
GO TO: www.scilinks.org
***sci*LINKS NUMBER:** HSTE345

2 Teach

MATH and MORE

Ask students how they would estimate wavelength and wave period if they were in a boat at sea. What visual marks might they use? Make sure students understand that wave period and wave speed are inversely proportional. That is, for a given wavelength, an increase in wave period means a decrease in wave speed. Have students work through several examples in their ScienceLog to see the relationships between the three wave measurements:

- The value of the wavelength is greater than the value of the wave period. (speed is greater than 1 m/s)
- The value of the wavelength is equal to the value of the wave period. (speed is equal to 1 m/s)
- The value of the wavelength is less than the value of the wave period. (speed is less than 1 m/s)

Answer to MATHBREAK

The wave speed is 0.6 m/s.

Teaching Transparency 159 "Wave Period"

Directed Reading Worksheet Section 2

2 Teach, *continued*

READING STRATEGY

Activity Have students reproduce **Figures 12** and **13** in their ScienceLog. As you read the section, pause at each term, and have them label the terms in their diagram. Make sure students understand each new term before proceeding.
Sheltered English

BRAIN FOOD

Ask the class why sighting a line of offshore breakers might cause sailors to consider turning their boats around. (As water becomes shallower, the wave height increases and the waves may break. Breaking waves could signal a submerged sandbar or reef; a ship could run aground if it encounters such an obstacle.)

REAL-WORLD CONNECTION

Discuss with students what they should do if they are ever caught in an undertow. Explain that instead of trying to swim against the current, they should signal for help and swim parallel to the shore. This will get them out of the undertow. They can then swim to shore. Diagram this scenario for students, and have them show you the proper direction to swim.

Teaching Transparency 160
"Deep-Water Waves Become Shallow-Water Waves"

Deep-Water Waves and Shallow-Water Waves Have you ever wondered why waves increase in height as they approach the shore? The answer has to do with the depth of the water. *Deep-water waves* are waves that move in water that is deeper than one-half of their wavelength. When the waves reach water that is shallower than one-half of their wavelength, they begin to interact with the ocean floor. These waves are called *shallow-water waves*. **Figure 12** shows how deep-water waves become shallow-water waves as they move toward the shore.

As deep-water waves become shallow-water waves, the water particles slow down and build up. This forces more water between wave crests and increases wave height. Gravity eventually pulls the high wave crests down, causing them to crash into the ocean floor as **breakers.** The area where waves first begin to tumble downward, or break, is called the *breaker zone*. Waves continue to break as they move from the breaker zone to the shore. The area between the breaker zone and the shore is called the **surf.**

Figure 12 *Deep-water waves become shallow-water waves when they reach depths of less than half of their wavelength.*

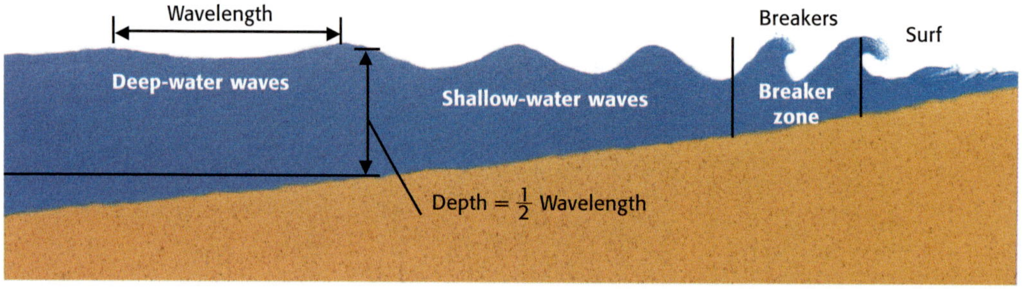

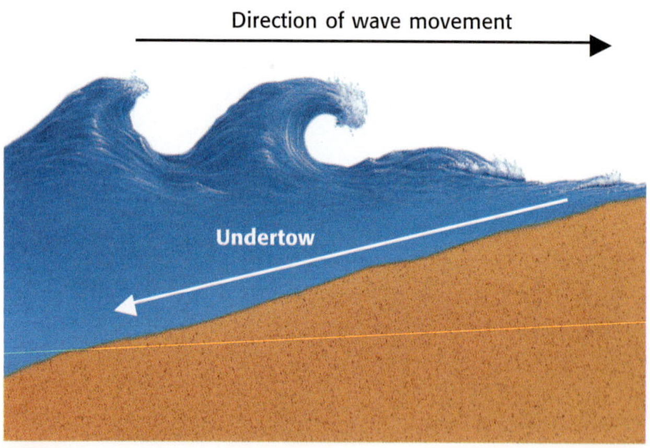

When waves crash on the beach head-on, the water they moved through flows back to the ocean underneath new incoming waves. This movement of water, which carries sand, rock particles, and plankton away from the shore, is called an *undertow*. **Figure 13** illustrates the back-and-forth movement of water at the shore.

Figure 13 *Head-on waves create an undertow.*

80

SCIENCE HUMOR

Q: What do two oceans say when they meet?

A: long time no sea

When waves hit the shore at an angle, they cause water to move along the shore in a current called a *longshore current*. This process is shown in **Figure 14.** Longshore currents are responsible for most sediment transport in beach environments. This movement of sand and other sediment both tears down and builds up the coastline. Unfortunately, longshore currents also carry trash and other types of ocean pollution, spreading it along the shore.

Figure 14 *Longshore currents form where waves approach beaches at an angle.*

Open-Ocean Waves Sometimes waves called whitecaps form in the open ocean. **Whitecaps** are white, foaming waves with very steep crests that break in the open ocean before the waves get close to the shore. These waves usually form during stormy weather, and they are usually short-lived. Calmer winds form waves called swells. **Swells** are rolling waves that move in a steady procession across the ocean. Swells have longer wavelengths than whitecaps and can travel for thousands of kilometers. **Figure 15** shows how whitecaps and swells differ.

Figure 15 *Whitecaps, shown in the photo at left, break in the open ocean, while swells, shown in the photo at right, roll gently in the open ocean.*

MATH and MORE

Whitecaps are usually caused by strong winds. When the wind blows more than 13 km/h, wave height increases faster than wavelength. When wave height is more than one-seventh of the wavelength, whitecaps form. If the wavelength is 3 m, what is the minimum wave height for whitecaps to form? (about 43 cm)

If the wavelength is 10 m, what is the minimum wave height for whitecaps to form? (about 1.43 m)

If whitecaps begin to form when the wave height reaches 4 m, what is the wavelength? (about 28 m)

Math Skills Worksheet "What Is a Fraction?"

Homework

Writing Surfing originated in the South Seas about 2,500 years ago. Polynesian sailors who couldn't get their boats through the rough waves near the shore would surf to land. It is now a recreational sport practiced all over the world.

Have students write a report about the history of surfing. Their report should include an explanation of the characteristics that make specific locations around the world ideal for the sport.

CONNECT TO ENVIRONMENTAL SCIENCE

Beach nourishment involves dredging enormous amounts of sand from the ocean floor and pumping it back onto the shoreline, rebuilding beaches that have been eroded by long-shore currents. It is a common, but controversial, practice along the Atlantic Coast. Some people think the natural erosion of beaches should not be disrupted. Beach erosion is influenced by other human activities, however. For example, artificial dams prevent sediment from reaching protective barrier islands and the dredging of ship channels disrupts the natural movement of sediment along the coast. Encourage students to investigate this issue and form their own opinions.

3 Extend

CONNECT TO METEOROLOGY

The sciences of oceanography and meteorology are combined in the duties of the National Oceanic and Atmospheric Administration (NOAA). Founded in 1970, this federal agency forecasts weather and monitors potentially destructive natural events, such as hurricanes, floods, and tsunamis. Interested students can write to NOAA or visit their Web site for more information.

GOING FURTHER

Writing | Have students research a tsunami that occurred in recent history and write a brief newspaper-style article about it. The articles should explain the cause and effects of the tsunami. Students could learn about the tsunami that struck the Pacific coast of Nicaragua in 1992 or the one that devastated Papua New Guinea in 1998. Encourage students to read their articles to the class.

Answers to QuickLab

4. Answers will vary. As wave energy passed through the rope, the ribbon moved up and down, but it did not move closer to the doorknob. Similarly, wave energy passes through water, moving it up and down, but it does not transport water.
6. Answers will vary. The wave height increased, but wavelength decreased; more waves occurred along the same length of rope.

QuickLab

Do the Wave

1. Tie one end of a thin piece of **rope** to a doorknob.
2. Tie a **ribbon** around the rope halfway between the doorknob and the other end of the rope.
3. Holding the rope at the untied end, quickly move the rope up and down, and observe the ribbon.
4. How does the movement of the rope and ribbon relate to the movement of water and deep-water waves?
5. Repeat step 3, but move the rope higher and lower this time.
6. How does this affect the waves in the rope?

TRY at HOME

Tsunamis Professional surfers often travel to Hawaii to catch some of the highest waves in the world. But even the best surfers would not be able to handle a tsunami. **Tsunamis** are waves that form when a large volume of ocean water is suddenly moved up or down. This movement can be caused by underwater earthquakes, volcanic eruptions, landslides, underwater explosions, or the impact of a meteorite or comet. The majority of tsunamis occur in the Pacific Ocean because of the greater number of earthquakes in that region. **Figure 16** shows how an earthquake can generate a tsunami.

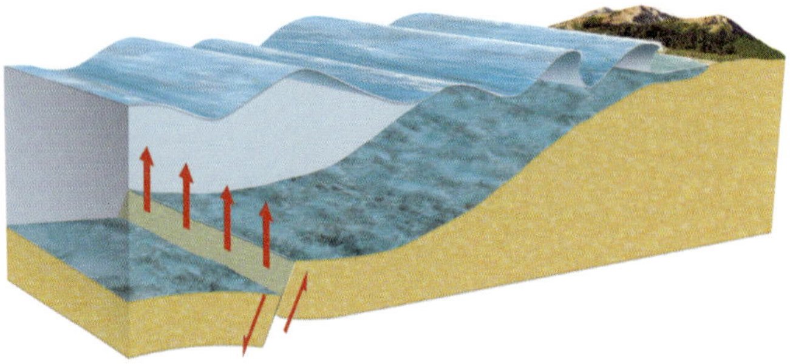

Figure 16 *An upward shift in the ocean floor creates an earthquake. The energy released by the earthquake pushes a large volume of water upward, creating a series of tsunamis.*

When tsunamis near continents, they slow down and their wavelengths shorten as they interact with the ocean floor. As tsunamis get closer together, their wave height increases. Tsunamis can reach more than 30 m in height as they slam into the coast, destroying just about everything in their path. The powerful undertow created by a tsunami can be as destructive as the tsunami itself. **Figure 17** shows a coastal community devastated by a tsunami.

Figure 17 *Imagine the strength of the tsunami that carried this boat so far inland!*

82

WEIRD SCIENCE

In 1946, the crew of a freighter anchored offshore near Hilo, Hawaii, were astounded to witness an enormous tsunami crash on the shore, crushing buildings, felling trees, and carrying boats, piers, and rocks ashore. Moments before, the wave had passed beneath the freighter unnoticed!

IS THAT A FACT!

The highest recorded tsunami was 64 m (210 ft) high; it struck Kamchatka, Siberia, in 1737.

APPLY

Timing a Tsunami

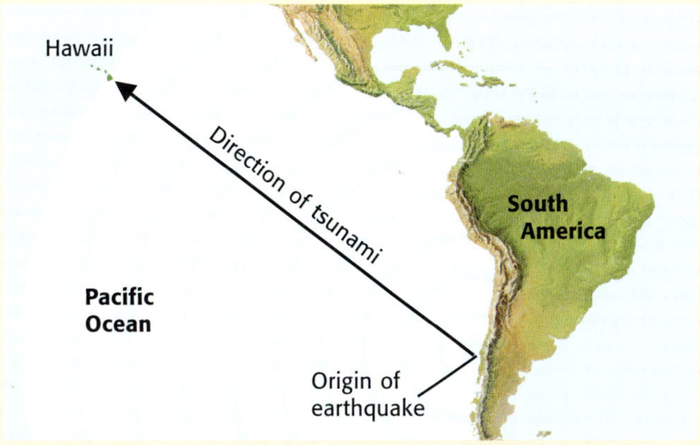

On May 22, 1960, an earthquake off the coast of South America generated a tsunami that completely crossed the Pacific Ocean. Ten thousand kilometers away from the origin of the earthquake, the tsunami hit the city of Hilo on the coast of Hawaii, causing extensive damage.

If the tsunami traveled at a speed of 188 m/s, how long after the earthquake occurred did the tsunami reach Hilo? If the residents of Hilo heard about the earthquake as soon as it happened, do you think they had enough warning time? What might be done to ensure that this amount of time would be sufficient warning for a tsunami?

Storm Surges

A **storm surge** is a local rise in sea level near the shore that is caused by strong winds from a storm, such as a hurricane. Winds form a storm surge by blowing water into a big pile under the storm. As the storm moves onto shore, so does the giant mass of water beneath it. Storm surges often disappear as quickly as they form, making them difficult to study. Storm surges contain a lot of energy and can reach about 8 m in height. This often makes them the most destructive part of hurricanes.

SECTION REVIEW

1. Explain how water moves as waves travel through it.
2. Where do deep-water waves become shallow-water waves?
3. Name five events that can cause a tsunami.
4. **Doing Calculations** Look again at Figure 11. If the wave speed is 0.8 m/s, what is the wavelength?

internet**connect**

SCI LINKS NSTA

TOPIC: Ocean Waves
GO TO: www.scilinks.org
***sci*LINKS NUMBER:** HSTE340

Answers to Section Review

1. Water at and near the surface rises and falls in circular movements as waves move through the water.
2. Deep-water waves become shallow-water waves in water that is shallower than one-half the wavelength of the deep-water waves.
3. Five events that can cause a tsunami are underwater earthquakes, volcanic eruptions, landslides, underwater explosions, and the impact of a cosmic body.
4. The wavelength is 4 m.

4 Close

Answers to APPLY

The tsunami reached Hilo almost 15 hours after the earthquake occurred. Fifteen hours may seem like a long time, but evacuating people from a coastal area is a major task. It is important for people who live along the coast to be educated about tsunamis. To better prepare for tsunamis, people could make an evacuation plan, perform evacuation drills, and set aside nonperishable food and medical supplies for emergency use.

Quiz

Concept Mapping Have students create a concept map linking section concepts and vocabulary.

Alternative Assessment

PORTFOLIO Using the wave characteristics described in this section, have students construct diagrams of different types of waves with varied characteristics. Use the transparency below to review wave characteristics with students. The diagrams should be made on poster board so they can be shared with the class and displayed in the classroom.

Teaching Transparency 282
"Wave Speed, Wavelength, and Frequency"
LINK TO PHYSICAL SCIENCE

Reinforcement Worksheet
"Waves to Your Pen Pal"

Section 3

Focus

Tides

The gravitational attraction of the moon and the sun creates tides on Earth. Students will learn how the position of Earth in relation to the moon and the sun creates different kinds of tides. They will also learn about the effects of coastal topography on tidewaters.

Bellringer

Measure 0.75 m toward the ceiling from a spot in the center of a wall, and mark this point with a piece of tape. Beside the measurement, write "average high tide." Then measure 0.75 m below the center spot, and label it "average low tide." Ask students, "What do you think is the greatest tidal range in the world?"

After students have guessed, tell them that the world's greatest tidal range occurs in the Bay of Fundy, between Nova Scotia and Maine. There, the tidal range can be more than 15 m. Be prepared to show students how high 15 m (or 10 times the example on the wall) is. Sheltered English

1) Motivate

Activity

If possible, plan a visit to a coastal area. Beforehand, have the class research intertidal zones and the organisms that inhabit these areas. If a visit to the coast is not possible, have students research intertidal organisms and their survival strategies and share their findings with the class.

Section 3 Reading Warm-Up

Terms to Learn

tides
tidal range
spring tides
neap tides

What You'll Do

- Explain tides and their relationship with the Earth, the sun, and the moon.
- Classify different types of tides.
- Analyze the relationship between tides and coastal land.

Tides

You have learned how winds and earthquakes can move ocean water. But there are less-obvious forces that continually move ocean water in regular patterns called tides. **Tides** are daily movements of ocean water that change the level of the ocean's surface. Tides are influenced by the sun and the moon, and they occur in a variety of cycles.

The Lure of the Moon

The phases of the moon and their relationship to the tides were first discovered more than 2,000 years ago by a Greek explorer named Pytheas. But Pytheas and other early investigators could not explain the relationship. A scientific explanation was not given until 1687, when Sir Isaac Newton's theories on the principle of gravitational pull were published. The gravity of the moon pulls on every particle of the Earth, but the pull is much more noticeable in liquids than in solids. This is because liquids move more easily. Even the liquid in an open soft drink is slightly pulled by the moon's gravity.

Gravitational forces from both the sun and the moon continuously pull on the Earth. Although the moon is much smaller than the sun, the moon's gravity is the dominant force behind Earth's tides.

High Tide and Low Tide How high tides get and how often they occur depend on the position of the moon as it revolves around the Earth. The moon's pull is strongest on the part of the Earth directly facing the moon. When that part happens to be a part of the ocean, the water there bulges toward the moon.

84

Misconception Alert

Students may think that water flows horizontally toward or away from the shore due to tides. As students read this section, make sure they understand that low tide occurs in an area because the water is actually pulled away from the area. This decreases the volume of water in that location and causes the water level to lower and the shoreline to move seaward.

At the same time, water on the opposite side of the Earth bulges due to the motion of the Earth and the moon around each other. These bulges are called *high tides*. Notice in **Figure 18** how the position of the moon causes the water to bulge. Also notice that when high tides occur, water is drawn away from the area between the high tides, causing *low tides* to form.

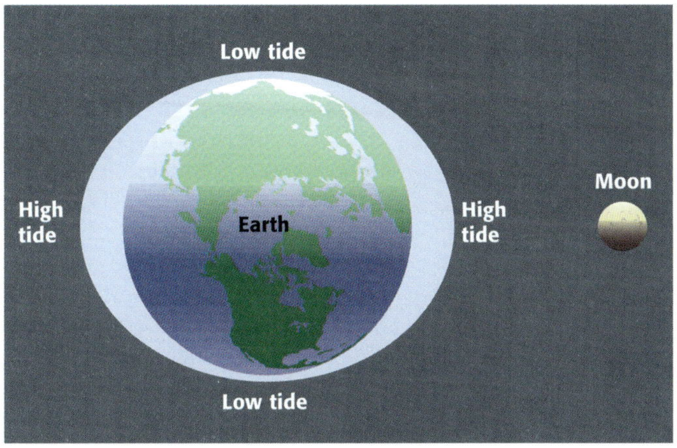

Figure 18 *High tide occurs on the part of Earth that is closest to the moon. At the same time, high tide also occurs on the opposite side of Earth.*

Timing the Tides The rotation of the Earth and the moon's revolution around the Earth determine when tides occur. If the Earth rotated at the same speed that the moon revolves around the Earth, tides would continuously occur at the same spots on Earth. But the moon revolves around the Earth much more slowly than the Earth rotates. **Figure 19** shows that it takes 24 hours and 50 minutes for a spot on Earth that is facing the moon to rotate so that it is facing the moon again.

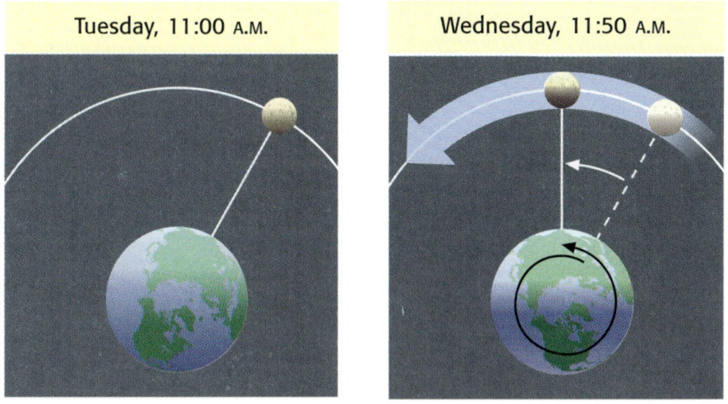

Figure 19 *Tides occur at different spots on Earth because the Earth rotates more quickly than the moon revolves around the Earth.*

LabBook

Puzzled about why high tide also occurs on the side of the Earth opposite the moon? Turn to page 106 to see how you can find out for yourself.

BRAIN FOOD

Even dry land has tides. For example, the land in Oklahoma moves up and down several centimeters throughout the day, corresponding with the tides. Tides on the solid part of Earth's surface are usually about one-third the size of ocean tides.

2 Teach

LabBook PG 106
Turning the Tides

USING THE FIGURE

Remind students that the Earth rotates at a much faster speed than the moon orbits. In **Figure 19,** point out that the distance the moon traveled is much shorter than the distance the spot on Earth facing the moon traveled. The spot on the Earth made more than a full rotation around the Earth's axis.

RETEACHING

Help students understand the hypothetical scenario discussed in the text (the moon revolving around the Earth at the same speed Earth rotates). Have students imagine the moon attached to a pole that is anchored to the Earth. In this scenario, the moon would turn with the Earth as the Earth rotates. The moon would always be facing the same spot on the Earth, so high tide would always occur at that spot and at the spot on the opposite side of the Earth. Make sure students realize this is not the case and that because the moon's revolution and the Earth's rotation occur at different speeds, the location of tides varies constantly.

SCIENTISTS AT ODDS

The cause of tides was a hotly debated topic in the sixteenth and seventeenth centuries. Galileo suggested that there was a connection between tides and the Earth's motion. To Galileo, tides proved beyond all doubt that Earth was moving. He argued that because Earth's waters are moving, Earth must be moving too. Johannes Kepler, another prominent scientist of the time, argued that the tides were linked to the moon's phases. Galileo made such a convincing argument that Kepler's ideas were dismissed. It was not until after Newton that scientists accepted that the gravitational forces exerted by the moon and the sun as well as the Earth's rotation are responsible for the tides.

Directed Reading Worksheet Section 3

3 Extend

MEETING INDIVIDUAL NEEDS

Learners Having Difficulty
Use this demonstration to show how tides are caused.

1. Use a small ball to represent the moon, a globe to represent Earth, and a large ball to represent the sun.
2. Ask a volunteer to help you show how the moon and Earth revolve around the sun.
3. Remind students that the moon and the sun are both "pulling" on Earth.
4. Point out that the moon's "pull" is greater than the sun's. Regions on Earth that are in a straight line from the moon experience high tides.
5. Point out that regions on Earth that face 90° away from the moon experience low tides.

Have students repeat the demonstration for each other to strengthen their understanding. Then ask students to read the text that explains spring tides and neap tides. Challenge students to use the globe and balls to demonstrate how spring tides and neap tides are caused.
Sheltered English

Homework

Graphing Help students look in the newspaper or on the Internet to find daily tidal information for a certain area for 1 month. Record the high- and low-tide measurements on a large chart that can be displayed in the classroom. At the end of the month, determine when spring tide and neap tide occurred. Compare these dates with the full, new, and quarter moon dates on a calendar, and have students graph this data to explain why the dates correspond.

Tidal Variations

The sun also affects tides. The sun is much larger than the moon, but it is also much farther away. As a result, the sun's influence on tides is less powerful than the moon's influence. The combined forces of the sun and the moon on the Earth result in tidal ranges that vary based on the positions of all three bodies. A **tidal range** is the difference between levels of ocean water at high tide and low tide.

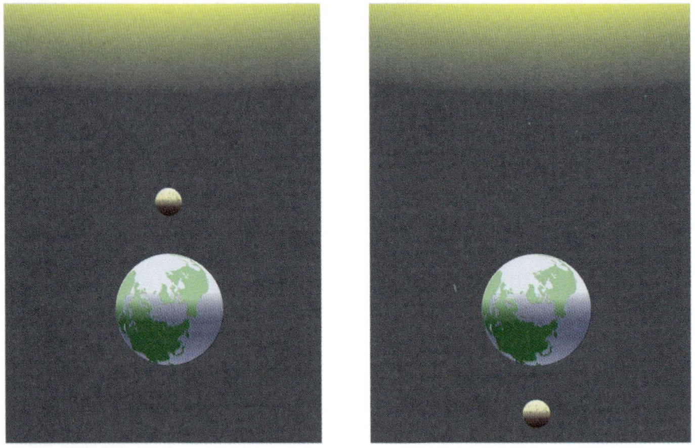

Figure 20 *During spring tides, the gravitational forces of the sun and moon pull on the Earth either from the same direction (left) or from opposite directions (right).*

Spring Tides When the sun, Earth, and moon are in alignment with one another, spring tides occur. **Spring tides** are tides with maximum daily tidal range that occur during the new and full moons. Spring tides occur every 14 days. The first time spring tides occur is when the moon is between the sun and Earth. The second time spring tides occur is when the moon and the sun are on opposite sides of the Earth. **Figure 20** shows the positions of the sun and moon during spring tides.

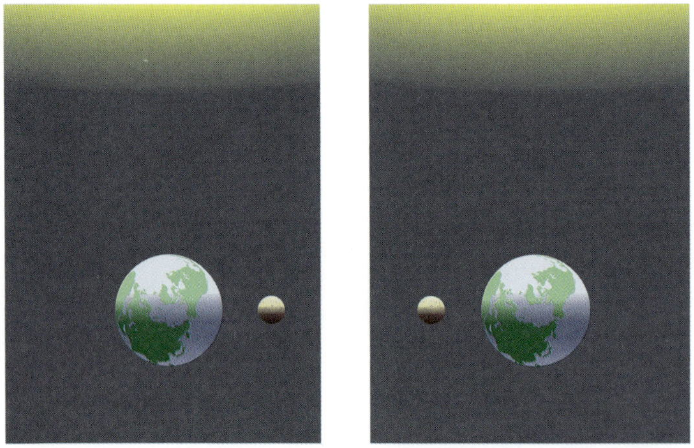

Figure 21 *During neap tides, the sun and moon are at right angles with respect to the Earth. This arrangement minimizes their gravitational effect on the Earth.*

Neap Tides When the sun, Earth, and moon form a 90° angle, neap tides occur. **Neap tides** are tides with minimum daily tidal range that occur during the first and third quarters of the moon. Neap tides occur halfway between the occurrence of spring tides. When neap tides occur, the gravitational forces on the Earth by the sun and the moon work against each other. **Figure 21** shows the positions of the sun and moon during neap tides.

86

Teaching Transparency 161
"Tidal Variations"

internetconnect
SciLINKS NSTA
TOPIC: The Tides
GO TO: www.scilinks.org
***sci*LINKS NUMBER:** HSTE350

WEIRD SCIENCE

The moon also creates tides in our atmosphere called lunar winds. Lunar winds move eastward in the morning and westward in the evening. Although these tides travel only 0.08 km/h, they can be detected by studying slight fluctuations in weather patterns.

Tides and Topography

Tides can be accurately predicted once the tidal range has been measured at a certain point over a period of time. This information can be useful for people who live near or visit the coast, as illustrated in **Figure 22.**

Figure 22 *It's a good thing the people on the beach (left) knew when high tide occurred (right). These photos show the Bay of Fundy, in New Brunswick, Canada. The Bay of Fundy has the greatest tidal range on Earth.*

In some coastal areas with narrow inlets, movements of water called tidal bores occur. A *tidal bore* is a body of water that rushes up through a narrow bay, estuary, or river channel during the rise of high tide, causing a very sudden tidal rise. Sometimes tidal bores form waves that rush up the inlets. Tidal bores occur in coastal areas of China, the British Isles, France, and Canada.

SECTION REVIEW

1. At what two spots on Earth does high tide occur?
2. Which tides have minimum tidal range? Which tides have maximum tidal range?
3. What causes tidal bores?
4. **Applying Concepts** How many days pass between minimum and maximum tidal range in any given area? Explain.

internetconnect

SCILINKS NSTA

TOPIC: The Tides
GO TO: www.scilinks.org
***sci*LINKS NUMBER:** HSTE350

Answers to Section Review

1. High tide occurs at the spot on Earth directly facing the moon and at the spot on Earth directly opposite the spot facing the moon.
2. Neap tides have minimum tidal range, and spring tides have maximum tidal range.
3. Tidal bores are caused by high tide rising in a narrow coastal inlet.
4. Seven days pass between minimum and maximum tidal range in any given area. Spring tides occur every 14 days, and neap tides occur midway between the occurrence of spring tides. Neap tides also occur every 14 days. Therefore, midway between the cycle of each type of tide, the other type occurs. The two types of tides alternate every 7 days.

CONNECT TO ENVIRONMENTAL SCIENCE

Tidal energy is one of the oldest renewable resources used by human civilizations. Tidal mills were used in England and France more than 900 years ago. Today, there are tidal power plants on the coast of France; on the Barents Sea, in Russia; and at Royal, Nova Scotia, in Canada. Have students research and make models of tidal energy plants.

4 Close

Quiz

1. Why are tides more noticeable in Earth's oceans than on its land? (Because the oceans are liquid, they flow more easily than solid rock.)
2. Why do spring tides exhibit such extremes of range? (When the moon, the sun, and Earth are aligned, as they are during spring tides, the tidal pull is maximized.)

Alternative Assessment

PORTFOLIO

Draw a random configuration of the sun, the moon, and Earth. Have students copy the drawing onto a sheet of paper. Then ask them to draw the tidal bulges caused by that configuration. Finally, have students draw the positions of the sun, moon, and Earth during spring and neap tides.

Critical Thinking Worksheet "Tides of Trouble"

Reinforcement Worksheet "But What About the Tides?"

Up from the Depths
Teacher's Notes

Time Required
One 45-minute class period

Lab Ratings
EASY → HARD

Teacher Prep: 1
Student Set-Up: 2
Concept Level: 2
Clean Up: 2

MATERIALS

The materials listed on the student page are enough for a group of 4–5 students. Note that the food coloring is used only to distinguish the water layers. Any two colors will work. You may find it simpler to make a roll of plastic wrap available to the class. Groups can then take a piece when they reach the appropriate steps.

Safety Caution
Remind students to review all safety cautions and icons before beginning this lab activity.

Answers
13. The warm, red water remained on top of the cold, blue water. There was very little mixing (if any), and there was no turning over.
15. The cold, blue water did not remain on top of the warm, red water. There was very little mixing (if any), and the water turned over.

Skill Builder Lab

USING SCIENTIFIC METHODS

Up from the Depths

Every year, the water in certain parts of the ocean "turns over." That is, the water at the bottom rises to the top and the water at the top falls to the bottom. This yearly change brings fresh nutrients from the bottom of the ocean to the fish living near the surface. So, it is a great time for fishing! However, the water in some parts of the ocean never turns over. You will use this activity to find out why.

Some parts of the ocean are warmer at the bottom, and some are warmer at the top. Sometimes the saltiest water is at the bottom; sometimes it is not. You will investigate how these factors help determine whether the water will turn over.

MATERIALS

- 400 mL beakers (5)
- tap water
- blue and red food coloring
- spoon
- bucket of ice
- watch or clock
- hot plate
- heat-resistant gloves
- 4 pieces of plastic wrap, approximately 30 cm × 20 cm
- salt

Ask a Question

1. Why do some parts of the ocean turn over, while others do not?

Conduct an Experiment

2. Label the beakers 1 through 5. Fill beakers 1 through 4 with tap water.
3. Add a drop of blue food coloring to the water in beakers 1 and 2 and stir.
4. Place beaker 1 in the bucket of ice for 10 minutes.
5. Add a drop of red food coloring to the water in beakers 3 and 4 and stir.
6. Set beaker 3 on a hot plate turned to a low setting for 10 minutes.
7. Add one spoonful of salt to the water in beaker 4, and stir.
8. While beaker 1 is cooling and beaker 3 is heating, copy the data table on the next page into your ScienceLog.

Gordon Zibelman
Drexel Hill Middle School
Drexel Hill, Pennsylvania

Observations Chart	
Mixture of water	**Observations**
Warm water placed above cold water	
Cold water placed above warm water	
Salty water placed above fresh water	DO NOT WRITE IN BOOK
Fresh water placed above salty water	

9 Pour half of the water in beaker 1 into beaker 5. Return beaker 1 to the bucket of ice.

10 Tuck a sheet of plastic wrap into beaker 5 so that the plastic rests on the surface of the water and lines the upper half of the beaker.

11 Put on your gloves. Slowly pour half of the water in beaker 3 into the plastic-lined upper half of beaker 5 to form two layers of water. Return beaker 3 to the hot plate, and remove your gloves.

12 Very carefully pull on one edge of the plastic wrap and remove it so that the warm, red water rests on the cold, blue water. **CAUTION:** The plastic wrap may be warm.

Make Observations

13 Wait about 5 minutes, and then observe the layers in beaker 5. Did one layer remain on top of the other? Was there any mixing or turning over? Record your observations in your data table.

14 Empty beaker 5 and rinse it with clean tap water.

15 Repeat the procedure in steps 9–14. This time, pour warm, red water from beaker 3 on the bottom and cold, blue water from beaker 1 on top. (Use gloves when pouring warm water.)

16 Again repeat the procedure used in steps 9–14. This time pour blue tap water from beaker 2 on the bottom and red, salty water from beaker 4 on top.

17 Repeat the procedure used in steps 9–14 a third time. This time pour red, salty water from beaker 4 on the bottom and blue tap water from beaker 2 on top.

Analyze the Results

18 Compare the results of all four trials. Explain why the water turned over in some of the trials but not in all of them.

Draw Conclusions

19 What is the effect of temperature and salinity on the density of water?

20 What makes the temperature of ocean water decrease? What could make the salinity of ocean water increase?

21 What reasons can you give to explain why some parts of the ocean turn over in the spring while some do not?

Going Further

Suggest a method for setting up a model that tests the combined effects of temperature and salinity on the density of water. Consider using more than two water samples and dyes.

16. The red, salty water did not remain on top of the blue tap water. There was little mixing, and the water turned over.
17. The blue tap water remained on top of the red, salty water. There was little mixing, and the water did not turn over.
18. In each case, the denser water sank to the bottom. Cold water is denser than warm water. When put in a beaker with warm water, the cold water either stayed at the bottom or sank to the bottom. Salt water is denser than fresh water. When put in a beaker with fresh water, the salt water either stayed at the bottom or sank to the bottom.
19. The density of water increases as its temperature decreases—cold water is denser than warm water. The density of water increases as its salinity increases—salt water is denser than fresh water.
20. The temperature of water can decrease due to seasonal temperature fluctuations or cold wind blowing across the water's surface. Currents can also carry cooler water to an area. The salinity of water can increase when evaporation occurs or when ice forms on the water's surface. These processes leave salts behind and make the remaining water more dense.
21. Parts of the ocean that turn over do so because their density changes due to variations in salinity or temperature. Parts of the ocean that do not turn over must not experience significant variations in salinity or temperature.

Going Further

The following combinations could be used:

Top	Bottom
Salt/cold	Fresh/warm
Salt/warm	Fresh/cold
Fresh/cold	Salt/warm
Fresh/warm	Salt/cold

Datasheets for LabBook

Chapter Highlights

Vocabulary Definitions

Section 1

surface current a streamlike movement of water that occurs at or near the surface of the ocean

Coriolis effect the curving of moving objects from a straight path due to the Earth's rotation

deep current a streamlike movement of ocean water far below the surface

upwelling a process in which cold, nutrient-rich water from the deep ocean rises to the surface and replaces warm surface water

El Niño periodic change in the location of warm and cool surface waters in the Pacific Ocean

Section 2

crest the highest point of a wave

trough the lowest point of a wave

wavelength the distance between two adjacent wave crests or wave troughs

wave height the vertical distance between a wave's crest and its trough

wave period the time it takes for two wave crests (or troughs) to pass a fixed point

breaker a heightened water wave that begins to tumble downward, or break, upon nearing the shore

Chapter Highlights

Section 1

Vocabulary

surface current *(p. 71)*
Coriolis effect *(p. 72)*
deep current *(p. 74)*
upwelling *(p. 77)*
El Niño *(p. 77)*

Section Notes

- Currents are classified as surface currents and deep currents.
- Surface currents are controlled by three factors: global winds, the Coriolis effect, and continental deflections.
- Surface currents, such as the Gulf Stream, can be several thousand kilometers in length.
- Deep currents form where the density of ocean water increases. Water density depends on temperature and salinity.
- Surface currents affect the climate of the land near which they flow.

Section 2

Vocabulary

crest *(p. 78)*
trough *(p. 78)*
wavelength *(p. 78)*
wave height *(p. 78)*
wave period *(p. 79)*
breaker *(p. 80)*
surf *(p. 80)*
whitecap *(p. 81)*
swells *(p. 81)*
tsunami *(p. 82)*
storm surge *(p. 83)*

Section Notes

- Waves are made up of two main components—crests and troughs.
- Waves are usually created by the transfer of the wind's energy across the surface of the ocean.

☑ Skills Check

Math Concepts

TWO OUT OF THREE The wave equation on page 79 has three variables. If you know two of these variables, you can figure out the third. Take a look at the examples below.

1. wave speed = 0.6 m/s, wave period = 10 s
   wavelength = wave speed × wave period = 6 m

2. wave speed = 0.6 m/s, wavelength = 6 m

$$\text{wave period} = \frac{\text{wavelength}}{\text{wave speed}} = 10 \text{ s}$$

Visual Understanding

BREAKING WAVES Before shallow-water waves break, their wave height increases and their wavelength decreases. Look at Figure 12 on page 80 again. Notice that the waves are taller and that their crests are closer together near the breaker zone.

90

Lab and Activity Highlights

Up from the Depths PG 88

Turning the Tides PG 106

Datasheets for LabBook (blackline masters for these labs)

SECTION 2

- Waves travel through water near the water's surface, while the water itself rises and falls in circular movements.
- Waves travel in the direction the wind blows. If the wind blows over a long distance, the wavelength becomes very large and the waves travel quickly.
- Wind-generated waves are classified as deep-water and shallow-water waves.
- Tsunamis are dangerous waves that can be very destructive to coastal communities.

SECTION 3

Vocabulary

tides *(p. 84)*
tidal range *(p. 86)*
spring tides *(p. 86)*
neap tides *(p. 86)*

Section Notes

- Tides are caused by the gravitational forces of the moon and sun tugging on the Earth.
- The moon's gravity is the main force behind tides.
- The relative positions of the sun and moon with respect to Earth cause different tidal ranges.
- Maximum tidal range occurs during spring tides.
- Minimum tidal range occurs during neap tides.
- Tidal bores occur as high tide rises in narrow coastal inlets.

Labs

Turning the Tides *(p. 106)*

internet**connect**

go.hrw.com

GO TO: go.hrw.com

Visit the **HRW** Web site for a variety of learning tools related to this chapter. Just type in the keyword:

KEYWORD: HSTH20

SCILINKS NSTA

GO TO: www.scilinks.org

Visit the **National Science Teachers Association** on-line Web site for Internet resources related to this chapter. Just type in the ***sci*LINKS** number for more information about the topic:

TOPIC: Ocean Currents	***sci*LINKS NUMBER:** HSTE330
TOPIC: El Niño	***sci*LINKS NUMBER:** HSTE335
TOPIC: Ocean Waves	***sci*LINKS NUMBER:** HSTE340
TOPIC: Tsunamis	***sci*LINKS NUMBER:** HSTE345
TOPIC: The Tides	***sci*LINKS NUMBER:** HSTE350

91

VOCABULARY DEFINITIONS, *continued*

surf the area between the breaker zone and the shore

whitecap a white, foaming wave with a very steep crest that breaks in the open ocean before the wave gets close to the shore

swells rolling waves that move in a steady procession across the ocean

tsunami a wave that forms when a large volume of ocean water is suddenly moved up or down

storm surge a local rise in sea level near the shore that is caused by strong winds from a storm, such as a hurricane

SECTION 3

tides daily movements of ocean water that change the level of the ocean's surface

tidal range the difference between levels of ocean water at high tide and low tide

spring tides tides with maximum daily tidal range that occur during the new and full moons

neap tides tides with minimum daily tidal range that occur during the first and third quarters of the moon

Vocabulary Review Worksheet

Blackline masters of these Chapter Highlights can be found in the **Study Guide.**

Lab and Activity Highlights

LabBank

Whiz-Bang Demonstrations, Spin Cycle

Long-Term Projects & Research Ideas, An Ocean Commotion

Interactive Explorations CD-ROM

CD 3, Exploration 4, "Latitude Attitude"

Chapter Review Answers

Using Vocabulary

1. A wavelength is the distance between two adjacent wave crests or wave troughs, while wave height is the vertical distance between a wave's crest and its trough.
2. Whitecaps are waves that break in the open ocean, while swells are waves that roll gently.
3. A tsunami is a wave created when a large volume of ocean water is suddenly moved up or down, while a storm surge is a local rise in sea level near the shore that is caused by winds from a storm.
4. Spring tides occur when the sun, the Earth, and the moon are aligned; neap tides occur when the sun, the Earth, and the moon form a 90 degree angle. Tidal range is greatest during spring tide and least during neap tide.
5. Surface currents are directly controlled by wind.
6. El Niño reduces upwelling along the coast of South America.
7. Spring tides occur when the moon is between the Earth and the sun.
8. Tidal range is the difference between levels of ocean water at high tide and low tide.

Understanding Concepts

Multiple Choice

9. c
10. d
11. d
12. b
13. c
14. a

Chapter Review

USING VOCABULARY

For each pair of terms, explain the difference in their meaning.

1. wavelength/wave height
2. whitecap/swell
3. tsunami/storm surge
4. spring tide/neap tide

Replace the incorrect term in each of the following sentences with the correct term provided in the word bank below:

5. Deep currents are directly controlled by wind.
6. The Coriolis effect reduces upwelling along the coast of South America.
7. Neap tides occur when the moon is between the Earth and the sun.
8. A tidal bore is the difference between levels of ocean water at high tide and low tide.

Word bank: breakers, spring tides, tsunamis, surface currents, tidal range, El Niño.

UNDERSTANDING CONCEPTS

Multiple Choice

9. Surface currents are formed by
   a. the moon's gravity.
   b. the sun's gravity.
   c. wind.
   d. increased water density.

10. Deep currents form when
    a. cold air decreases water density.
    b. warm air increases water density.
    c. the ocean surface freezes and solids from the water underneath are removed.
    d. salinity increases.

11. When waves come near the shore,
    a. they speed up.
    b. they maintain their speed.
    c. their wavelength increases.
    d. their wave height increases.

12. Longshore currents transport sediment
    a. out to the open ocean.
    b. along the shore.
    c. during low tide only.
    d. during high tide only.

13. Whitecaps break
    a. in the surf.
    b. in the breaker zone.
    c. in the open ocean.
    d. as their wavelength increases.

14. Tidal range is greatest during
    a. spring tide.
    b. neap tide.
    c. a tidal bore.
    d. the day only.

92

Concept Mapping Transparency 14

Blackline masters of this Chapter Review can be found in the **Study Guide.**

Short Answer

15. Explain the relationship between upwelling and El Niño.

16. Explain what happens when the North Atlantic Deep Water meets the Antarctic Bottom Water.

17. Describe the relative positions of the Earth, the moon, and the sun during neap tide. Where do high tide and low tide occur during this time?

18. Explain the difference between the breaker zone and the surf.

Concept Mapping

19. Use the following terms to create a concept map: wind, deep currents, sun's gravity, types of ocean-water movement, surface currents, tides, increasing water density, waves, moon's gravity.

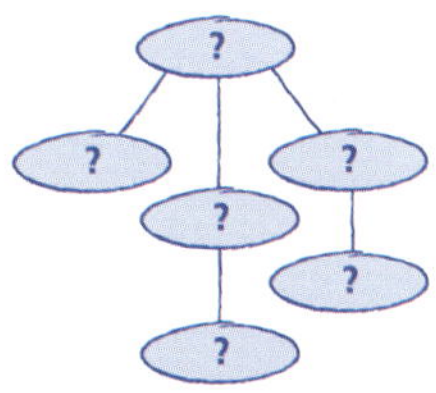

CRITICAL THINKING AND PROBLEM SOLVING

Write one or two sentences to answer the following questions:

20. What would happen to surface currents if the Earth reversed its rotation? Be specific.

21. How would you explain a bottle moving across the water in the same direction the waves are traveling?

22. You and a friend are planning a fishing trip to the ocean. Your friend tells you that the fish bite more in his secret fishing spot during low tide. If low tide occurred at the spot at 7 A.M. today and you are going to fish there in one week, at what time will low tide occur in that spot?

MATH IN SCIENCE

23. If a barrier island that is 1 km wide and 10 km long loses 1.5 m of its width per year to erosion by longshore current, how long will it take for the island to lose one-fourth of its width?

INTERPRETING GRAPHICS

Study the diagram below, and answer the questions that follow.

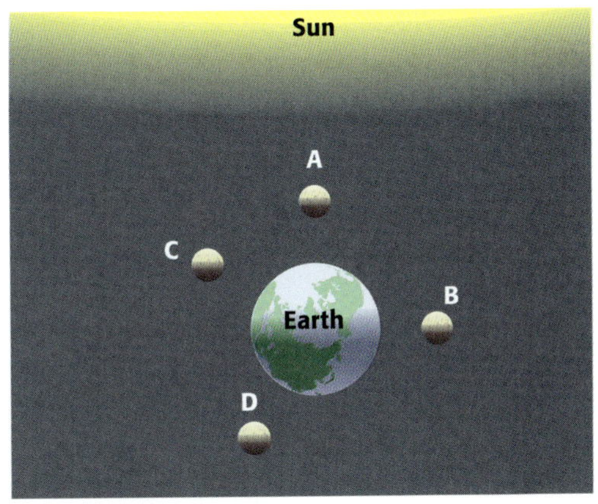

24. At which position (**A**, **B**, **C**, or **D**) would the moon be during a neap tide?

25. At which position (**A**, **B**, **C**, or **D**) would the moon be during a spring tide?

26. Would tidal range be greater with the moon at position **C** or position **D**? Why?

Reading Check-up

Take a minute to review your answers to the Pre-Reading Questions found at the bottom of page 68. Have your answers changed? If necessary, revise your answers based on what you have learned since you began this chapter.

Short Answer

15. When El Niño occurs, warm surface water remains along the Pacific coast of South America. This prevents upwelling from occurring along the coast.
16. The North Atlantic Deep Water flows on top of the Antarctic Bottom Water.
17. During neap tide, the sun, the moon, and the Earth form a right angle, with the Earth in the middle. High tide occurs on the side of the Earth facing the moon and on the side opposite the moon. Low tide occurs on the side of the Earth facing the sun and on the side opposite the sun.

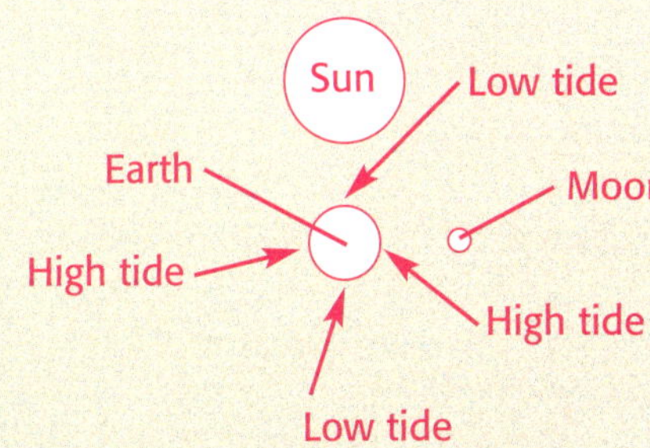

18. The breaker zone is where waves first begin to tumble downward. The surf is the zone between the breaker zone and the shore. In the surf, water moves toward the shore.

Concept Mapping

19. 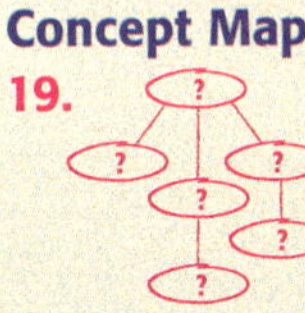 An answer to this exercise can be found in the front of this book.

Critical Thinking and Problem Solving

20. Surface currents in the Northern Hemisphere would rotate counterclockwise, and surface currents in the Southern Hemisphere would rotate clockwise.
21. There would have to be a surface current moving in the same direction as the waves are traveling.
22. 12:50 P.M. (1:15 A.M. is also acceptable.)

Math in Science

23. about 167 years

Interpreting Graphics

24. B
25. A
26. D; when the Earth-moon system is viewed from above the Earth's North Pole, the path of the moon around the Earth appears counterclockwise. In this case, the moon would be very near a spring-tide position at position D and very near a neap-tide position at position C.

CAREERS
Seismologist—Hiroo Kanamori

Background
Seismologists study earthquakes. Then why don't we call them earthquake-ologists? Well, the word *seismology* comes from the Greek language, and in Greek, *seismos* means "to shake"!

Dr. Kanamori recently received the Bucher Medal for his outstanding achievements in seismology. He bridged the gap between seismology and physics by developing an earthquake scale called the "moment magnitude scale." It rates earthquakes by the minimum energy released and is consistent with the Richter scale.

CAREERS

SEISMOLOGIST

As a seismologist, **Hiroo Kanamori** studies how earthquakes occur and tries to reduce their impact on our society. He also analyzes the effects of earthquakes on oceans and how earthquakes cause tsunamis (tsoo NAH mes). He has discovered that even weak earthquakes can create tsunamis.

Because most tsunamis are caused by underwater earthquakes, scientists can monitor earthquakes to predict when and where a tsunami will hit land. But the predictions are not always accurate. Very weak earthquakes should not create powerful tsunamis, yet they do. Kanamori calls these special events *tsunami earthquakes,* and he has learned how to predict the size of the resulting tsunamis.

A tsunami can be more dangerous than an earthquake. When people feel the tremors created when the plates slide, they don't always realize that a large tsunami may be on the way. Because of this, people don't expect a tsunami and don't leave the area.

Measuring Tsunami Earthquakes
As tectonic plates grind against each other, they send out seismic waves. These waves travel through the earth's crust and can be recorded by a sensitive machine. But when the plates grind very slowly, only long period seismic waves are recorded. When Kanamori sees a long period wave, he knows that a tsunami will form.

"The speed of the average tsunami is about 800 km/h per hour, which is much slower than the speed of the long period waves at 15,000 km/h. So these special seismic waves arrive at distant recording stations much earlier than a tsunami," explains Kanamori. This important fact lets scientists like Kanamori warn people in the tsunami's path so they can leave the area.

An Interesting Career
Kanamori finds his work very rewarding. "It is always good to see how what we learned in the classroom can solve our real-life problems," he explains. "We can see how physics and mathematics work to explain seemingly complex natural events, such as earthquakes, volcanoes, and tsunamis."

A Challenge
▶ The depth of an ocean influences how fast a tsunami travels. To investigate, fill a 0.5 m long tub with 5 cm of water. Tap the tub. How long does it take for the wave to go back and forth? Add more water, and tap it again. Did the wave move faster or slower?

▶ *Monster waves are well-known in many communities along the Pacific coast.*

94

Answers to A Challenge
The wave moves faster as the water gets deeper. Ask students what happens to the speed of a tsunami wave as it approaches a shoreline. (The tsunami wave slows down as it approaches the shoreline.) Point out that as a tsunami wave slows down, it becomes taller.

Health Watch

Red Tides

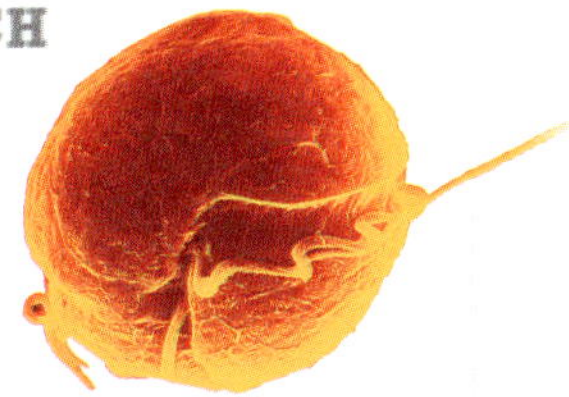

Imagine going to the beach only to find that the ocean water has turned red and fish are floating belly up all over the place. This is not an imaginary scene. It really happens. What could cause such widespread damage to the ocean? Single-cell algae, that's what!

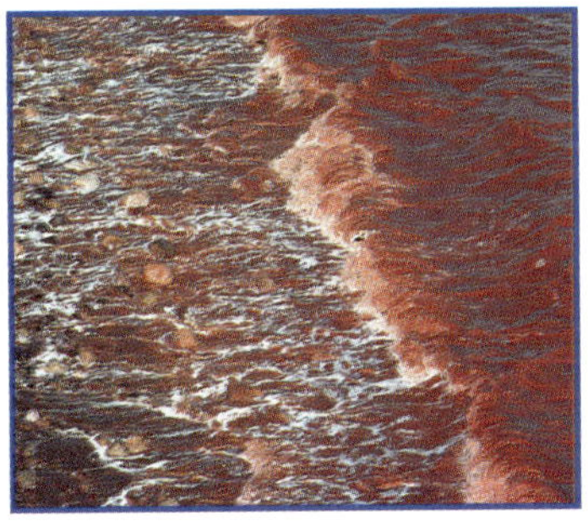

▲ *Harmful algal blooms are caused by algae like the one shown above right.*

Blooming Algae

When certain algae grow rapidly, they clump together on the ocean's surface in an algal bloom that changes the color of the water. People called these algal blooms red tides because the blooms often turned the water red or reddish-brown. They also believed that tidal conditions caused the blooms. Scientists now call these algae explosions harmful algal blooms (HABs) because HABs are not always red, and they are not directly related to tides. The blooms are harmful because certain species of algae produce toxins that can poison fish, shellfish, and people.

Scientists also have learned that the ocean's natural currents may carry HABs hundreds of miles along a coastline. For example, in 1987, the Gulf Stream off the Atlantic coast of Florida carried a toxic bloom up the coast to North Carolina.

Troublesome Toxins

Some people who ate tainted shellfish from the North Carolina coast in 1987 suffered from muscular aches, anxiety, sweating, dizziness, diarrhea, vomiting, and abdominal pain. Some algae toxins can even kill people who eat the tainted seafood. Another HAB occurred in 1987 in Nova Scotia, Canada. Four people died from eating contaminated shellfish, and another 150 people suffered from symptoms such as dizziness, headaches, seizures, short-term memory loss, and comas.

In the 1990s, Texas, Maryland, Alaska, and many other coastal states experienced HABs. However, the problem is not confined to North America. Throughout the 1990s, HABs caused health problems in South Africa, Argentina, India, New Zealand, and France.

No Signs to Read

Fish and shellfish are major sources of protein for people all over the world. Unfortunately, there are no outward signs when seafood is contaminated. The toxins don't change the flavor, and cooking the seafood doesn't eliminate the toxins. Sometimes a HAB rides into an area on an ocean current, causing fish to die and people to become ill before authorities are aware of the problem.

Fortunately, scientists all over the world are working on ways to monitor and even predict HABs. As a result, people eventually may be able to eat fish and shellfish without worrying about toxic algae.

Find Out More

▶ Some people think that human activities are causing more HABs than occurred in the past. Other people disagree. Find out more about this issue, and have a class debate about the role humans play in creating HABs.

Health Watch

Red Tides

Background

Of the 4,400 phytoplankton species, a mere 50 to 60 can produce toxins. Even a small dose of the toxin of certain species can prove fatal.

The best way for consumers to reduce the chances of buying poisoned seafood is to shop at a reputable seafood store. Seafood purchased in this way has passed through a government-run system that monitors ocean-water quality and toxin levels. Supermarkets are another safe option, as is ordering seafood at a reputable restaurant.

Answers to Find Out More

At present, no definitive study links HABs to human activities; all information is still speculation. Activities considered to be either directly or indirectly related to the cause of HABs include overfishing, global warming, and ocean-nutrient fluctuations due to coastal development and water runoff. Some scientists suggest that the increase in reported HAB events might be explained by the existence of better detection methods and the fact that more people are witnessing the events. Perhaps students will be able to find some specific studies linking human activities to HABs that would serve to sway the debate in support of such arguments.

Exploring, inventing, and investigating are essential to the study of science. However, these activities can also be dangerous. To make sure that your experiments and explorations are safe, you must be aware of a variety of safety guidelines.

You have probably heard of the saying, "It is better to be safe than sorry." This is particularly true in a science classroom where experiments and explorations are being performed. Being uninformed and careless can result in serious injuries. Don't take chances with your own safety or with anyone else's.

Following are important guidelines for staying safe in the science classroom. Your teacher may also have safety guidelines and tips that are specific to your classroom and laboratory. Take the time to be safe.

Safety Rules!

Start Out Right

Always get your teacher's permission before attempting any laboratory exploration. Read the procedures carefully, and pay particular attention to safety information and caution statements. If you are unsure about what a safety symbol means, look it up or ask your teacher. You cannot be too careful when it comes to safety. If an accident does occur, inform your teacher immediately, regardless of how minor you think the accident is.

If you are instructed to note the odor of a substance, wave the fumes toward your nose with your hand. Never put your nose close to the source.

Safety Symbols

All of the experiments and investigations in this book and their related worksheets include important safety symbols to alert you to particular safety concerns. Become familiar with these symbols so that when you see them, you will know what they mean and what to do. It is important that you read this entire safety section to learn about specific dangers in the laboratory.

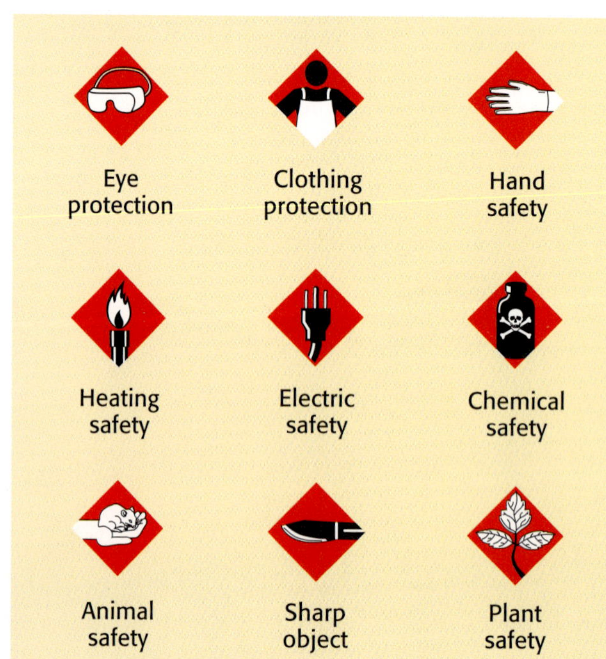

Eye Safety

Wear safety goggles when working around chemicals, acids, bases, or any type of flame or heating device. Wear safety goggles any time there is even the slightest chance that harm could come to your eyes. If any substance gets into your eyes, notify your teacher immediately, and flush your eyes with running water for at least 15 minutes. Treat any unknown chemical as if it were a dangerous chemical. Never look directly into the sun. Doing so could cause permanent blindness.

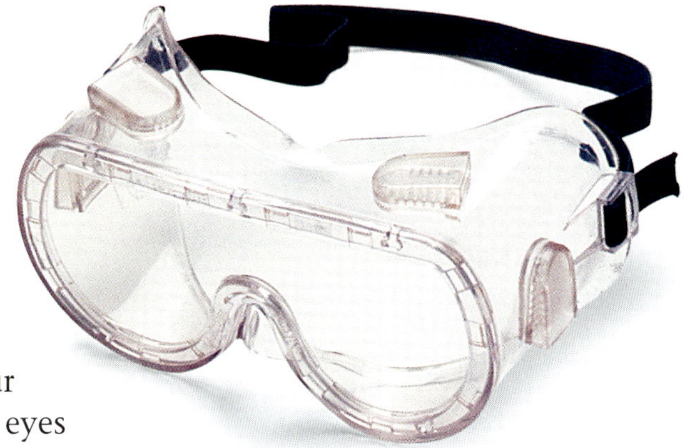

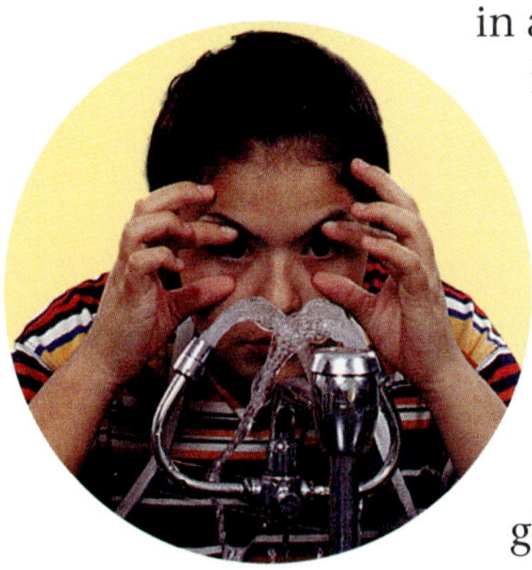

Avoid wearing contact lenses in a laboratory situation. Even if you are wearing safety goggles, chemicals can get between the contact lenses and your eyes. If your doctor requires that you wear contact lenses instead of glasses, wear eye-cup safety goggles in the lab.

Safety Equipment

Know the locations of the nearest fire alarms and any other safety equipment, such as fire blankets and eyewash fountains, as identified by your teacher, and know the procedures for using them.

Be extra careful when using any glassware. When adding a heavy object to a graduated cylinder, tilt the cylinder so the object slides slowly to the bottom.

Neatness

Keep your work area free of all unnecessary books and papers. Tie back long hair, and secure loose sleeves or other loose articles of clothing, such as ties and bows. Remove dangling jewelry. Don't wear open-toed shoes or sandals in the laboratory. Never eat, drink, or apply cosmetics in a laboratory setting. Food, drink, and cosmetics can easily become contaminated with dangerous materials.

Certain hair products (such as aerosol hair spray) are flammable and should not be worn while working near an open flame. Avoid wearing hair spray or hair gel on lab days.

Sharp/Pointed Objects

Use knives and other sharp instruments with extreme care. Never cut objects while holding them in your hands. Place objects on a suitable work surface for cutting.

Heat

Wear safety goggles when using a heating device or a flame. Whenever possible, use an electric hot plate as a heat source instead of an open flame. When heating materials in a test tube, always angle the test tube away from yourself and others. In order to avoid burns, wear heat-resistant gloves whenever instructed to do so.

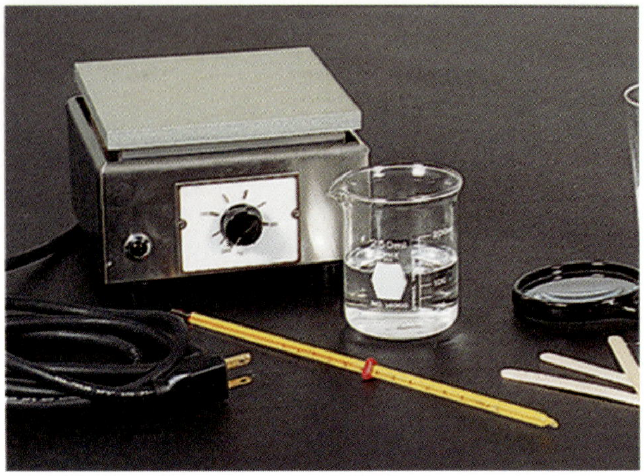

Chemicals

Wear safety goggles when handling any potentially dangerous chemicals, acids, or bases. If a chemical is unknown, handle it as you would a dangerous chemical. Wear an apron and safety gloves when working with acids or bases or whenever you are told to do so. If a spill gets on your skin or clothing, rinse it off immediately with water for at least 5 minutes while calling to your teacher.

Never mix chemicals unless your teacher tells you to do so. Never taste, touch, or smell chemicals unless you are specifically directed to do so. Before working with a flammable liquid or gas, check for the presence of any source of flame, spark, or heat.

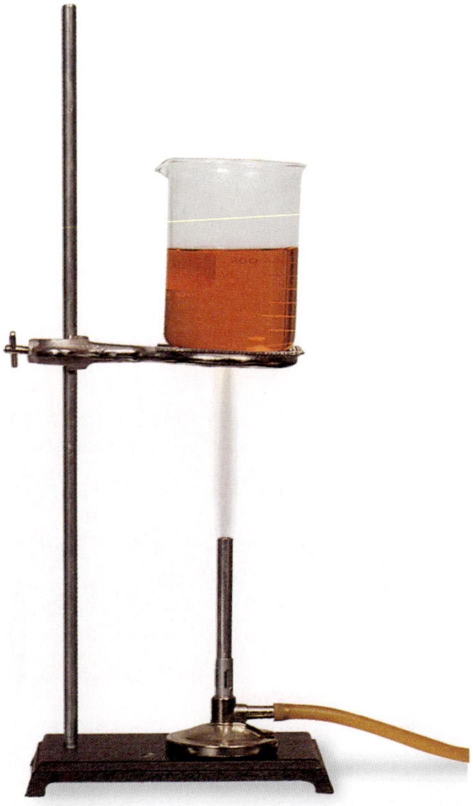

Electricity

Be careful with electrical cords. When using a microscope with a lamp, do not place the cord where it could trip someone. Do not let cords hang over a table edge in a way that could cause equipment to fall if the cord is accidentally pulled. Do not use equipment with damaged cords. Be sure your hands are dry and that the electrical equipment is in the "off" position before plugging it in. Turn off and unplug electrical equipment when you are finished.

Animal Safety

Always obtain your teacher's permission before bringing any animal into the school building. Handle animals only as your teacher directs. Always treat animals carefully and with respect. Wash your hands thoroughly after handling any animal.

Plant Safety

Do not eat any part of a plant or plant seed used in the laboratory. Wash hands thoroughly after handling any part of a plant. When in nature, do not pick any wild plants unless your teacher instructs you to do so.

Glassware

Examine all glassware before use. Be sure that glassware is clean and free of chips and cracks. Report damaged glassware to your teacher. Glass containers used for heating should be made of heat-resistant glass.

Clean Up Your Act
Teacher's Notes

Time Required
Two 45-minute class periods

Lab Rating
EASY → HARD

TEACHER PREP ▲▲▲
STUDENT SET-UP ▲▲▲
CONCEPT LEVEL ▲▲▲
CLEAN UP ▲▲▲

MATERIALS

The materials listed on the student page are enough for a group of 4–5 students. To keep results consistent with all lab groups, each group should have the same size gravel and the same size sand. Also, the layers of sand and gravel should be the same dimensions in each group.

Safety Caution
Remind students to review all safety cautions and icons before beginning this lab activity. Caution students not to taste any of the liquids in this lab.

Kenneth Creese
White Mountain Jr. High
Rock Springs, Wyoming

Clean Up Your Act

When you wash dishes, the family car, the bathroom sink, or your clothes, you wash them with water. But have you ever wondered how water gets clean? Two major methods of purifying water are filtration and evaporation. In this activity you will use both of these methods to test how well they remove pollutants from water. You will test detritus (decaying plant matter), soil, vinegar, and detergent. Your teacher may also ask you to test other pollutants.

Form a Hypothesis

1. Form a hypothesis about whether filtration and evaporation will clean each of the four pollutants from the water and how well they might do it. Then use the procedures below to test your hypothesis.

Part A: Filtration

Filtration is a common method of removing various pollutants from water. It requires very little energy—gravity pulls water down through the layers of filter material. See how well this energy-efficient method works to clean your sample of polluted water.

Conduct an Experiment

2. Put on your goggles. Use scissors to cut the bottom out of the empty soda bottle carefully.
3. Carefully punch four or five small holes through the plastic cap of the bottle using a small nail and hammer. Screw the plastic cap onto the bottle.
4. Turn the bottle upside down, and set its neck in a ring on a ring stand, as shown on the next page. Put a handful of gravel into the inverted bottle. Add a layer of activated charcoal, followed by thick layers of sand and gravel. Place a 400 mL beaker under the neck of the bottle.
5. Fill each of the large beakers with 1,000 mL of clean water. Set one beaker aside to serve as the control. Add three or four spoonfuls of each of the following pollutants to the other beaker: detritus, soil, household vinegar, and dishwashing detergent.

Materials

Part A
- scissors
- plastic 2 L soda bottle with cap
- hammer and small nail
- gravel
- activated charcoal
- sand

Part B
- Erlenmeyer flask
- one-hole rubber stopper with a glass tube
- 1.5 m of plastic tubing
- heat-resistant gloves
- hot plate
- sealable plastic sandwich bag
- ice

Parts A and B
- ring stand with ring
- 400 mL beaker
- 1,000 mL beakers (2)
- 2,000 mL of water
- detritus (grass and leaf clippings)
- soil
- household vinegar
- dishwashing detergent
- hand lens
- 2 plastic spoons
- pH test strips

Collect Data

6. Copy the table below into your ScienceLog, and record your observations for each beaker in the columns labeled "Before cleaning."
7. Observe the color of the water in each beaker.
8. Use a hand lens to examine the water for visible particles.
9. Smell the water, and note any unusual odors.
10. Stir the water in each beaker rapidly with a plastic spoon, and check for suds. Use a different spoon for each sample.
11. Use a pH test strip to find the pH of the water.
12. Gently stir the clean water, and then pour half of it through the filtration device.
13. Observe the water in the collection beaker for color, particles, odors, suds, and pH. Be patient. It may take several minutes for the water to travel through the filtration device.
14. Record your observations in the appropriate "After filtration" column in your table.
15. Repeat steps 12–14 using the polluted water.

Results Table						
	Before cleaning (clean water)	Before cleaning (polluted water)	After filtration (clean water)	After filtration (polluted water)	After evaporation (clean water)	After evaporation (polluted water)
Color						
Particles						
Odor						
Suds						
pH						

LabBook

Preparation Notes

Varying the thickness of the layers can contribute to a variation in the results. Specify the thickness of each layer so that the results between groups are comparable. The layers could be the following dimensions: 7 cm of gravel, 2.5 cm of charcoal, 10 cm of sand, and 10 cm of gravel. You may adjust the layers according to your class size or the size of the bottles.

For variation, try different sizes of gravel or different textures of sand. Both will affect how many particles travel through the filter. If you place a few drops of food coloring in the water, students can watch the progress of the water as it passes through the filter.

You may decide that it is easier and safer for you to perform step 2 of Part A, cutting the soda bottles, before class. You may also wish to perform steps 19–22 of Part B ahead of time. Some students may find it difficult to attach the plastic tubing to the glass tube or to slide the glass tube into the rubber stopper.

Datasheets for LabBook

Answers

16. The filtered water was lighter in color than the unfiltered water. The water was still not as clear as the clean water. The color of the clean water stayed about the same.
17. No; the filtration method did not remove all of the particles from the polluted water. Many of the particles passed through the filter, but there were fewer particles than before the filtration.
18. The pH of the water changed slightly. The final pH of the polluted water was not the same as the clean water. After the polluted water was filtered, its pH was still slightly more acidic than the clean water.

Analyze the Results

16. How did the color of the polluted water change after the filtration? Did the color of the clean water change?
17. Did the filtration method remove all of the particles from the polluted water? Explain.
18. How much did the pH of the polluted water change? Did the pH of the clean water change? Was the final pH of the polluted water the same as the pH of the clean water before cleaning? Explain.

Part B: Evaporation

Cleaning water by evaporation is more expensive than cleaning water by filtration. Evaporation requires more energy, which can come from a variety of sources. In this activity, you will use an electric hot plate as the energy source. See how well this method works to clean your sample of polluted water.

Conduct an Experiment

19. Fill an Erlenmeyer flask with about 250 mL of the clean water, and insert the rubber stopper and glass tube into the flask.
20. Wearing goggles and gloves, connect about 1.5 m of plastic tubing to the glass tube.
21. Set the flask on the hot plate, and run the plastic tubing up and around the ring and down into a clean, empty 400 mL collection beaker.
22. Fill the sandwich bag with ice, seal the bag, and place the bag on the ring stand. Be sure the plastic bag and the tubing touch, as shown below.
23. Bring the water in the flask to a slow boil. As the water vapor passes by the bag of ice, the vapor will condense and drip into the collection beaker.

Collect Data

24. Observe the water in the collection beaker for color, particles, odor, suds, and pH. Record your observations in the appropriate "After evaporation" column in your data table.

25. Repeat steps 23–24 using the polluted water.

Analyze the Results

26. How did the color of the polluted water change after evaporation? Did the color of the clean water change after evaporation?

27. Did the evaporation method remove all of the particles from the polluted water? Explain.

28. How much did the pH of the polluted water change? Did the pH of the final clean water change? Was the final pH of the polluted water the same as the pH of the clean water before it was cleaned? Explain.

Draw Conclusions (Parts A and B)

29. Which method—filtration or evaporation—removed the most pollutants from the water? Explain your reasoning.

30. Describe any changes that occurred in the clean water during this experiment.

31. What do you think are the advantages and disadvantages of each method?

32. Explain how you think each material (sand, gravel, and charcoal) used in the filtration system helped clean the water.

33. List areas of the country where you think each method of purification would be the most and the least beneficial. Explain your reasoning.

Going Further

Do you think either purification method would remove oil from water? If time permits, repeat your experiment using several spoonfuls of cooking oil as the pollutant.

Filtration is only one step in the purification of water at water-treatment plants. Research other methods used to purify public water supplies.

LabBook

Answers

26. After the evaporation process, the polluted water was clear. The color of the clean water did not change.

27. No particles were visible in the treated water after the evaporation method. The particles in the polluted water were left behind when the water evaporated.

28. Answers may vary, depending on the clean-water source. Generally, the pH of the two samples should be very close or the same.

29. The evaporation method removed the most pollutants from the water. The pollutants were left behind when the water evaporated.

30. The clean water picked up some particles as it traveled through the filter. It was not as clean after the filtration as it had been before the filtration.

31. The advantages of the filtration method include: it is easy to do; it can treat large amounts of water; and it works relatively quickly. The filtration method doesn't remove all of the pollutants, however.

    The evaporation method removed more of the pollutants, but it is time consuming and expensive, particularly with large volumes of water.

32. Answers will vary. The sand filtered out some of the larger particles and some of the soap. The gravel also removed the large particles. The charcoal removed most of the smaller particles, the odors, and some of the soap.

33. Answers will vary.

Going Further

- Oil is removed from the water in both methods. The filtration method is relatively quicker than the evaporation method.
- Other methods of water purification include: reverse osmosis; settling ponds; chemical additives, such as chlorine; and other filtration layers. Students may also find out about bioremediation and the use of bacteria and plants to clean polluted water.

Investigating an Oil Spill
Teacher's Notes

Time Required
One 45-minute class period

Lab Ratings
EASY → HARD

TEACHER PREP: 3
STUDENT SET-UP: 1
CONCEPT LEVEL: 2
CLEAN UP: 3

MATERIALS

The materials listed on the student page are sufficient for a group of 2–4 students. You may also choose to perform this activity as a demonstration to limit the use of machine oil.

Safety Caution
Remind students to review all safety cautions and icons before beginning this lab activity. Machine oil releases a strong odor. Use it only in well-ventilated areas.

Preparation Notes
Light machine oil may be replaced by cooking oil to demonstrate the same principle. If you choose to use machine oil, be sure there is sufficient ventilation in your classroom.

Disposal Information
Always follow federal, state, and local guidelines when disposing of oil. Pour cooking oil into a container of sand and put it in the trash.

DISCOVERY LAB

Investigating an Oil Spill

Have you ever wondered why it is important to bring used motor oil to a recycling center rather than simply pouring it down the nearest drain or sewer? Or have you ever wondered why an oil spill of only a few thousand liters into an ocean containing many millions of liters of water can cause so much damage? The reason has to do with the fact that a little oil goes a long way.

Materials
- safety gloves
- large pan (at least 22 cm in diameter)
- water
- pipet
- 15 mL light machine oil
- metric ruler
- graduated cylinder
- calculator (optional)

Observing Oil and Water
You may have heard the expression "Oil and water don't mix." This is true—oil dropped on water will spread out thinly over the surface of the water. In this activity, you'll learn exactly how far oil can spread when it is in contact with water.

Procedure
1. Fill the pan two-thirds full with water. Be sure to wear your goggles and gloves.
2. Using the pipet, carefully add one drop of oil to the water in the middle of the pan.
   Caution: Machine oil is poisonous. Keep materials that have contacted oil out of your mouth and eyes.
3. Observe what happens to the drop of oil for the next few seconds. Record your observations in your ScienceLog.
4. Using a metric ruler, measure the diameter of the oil slick to the nearest centimeter.
5. Determine the area of the oil slick in square centimeters by using the formula for finding the area of a circle ($A = \pi r^2$). The radius (r) is equal to the diameter you measured in step 4 divided by 2. Multiply the radius by itself to get the square of the radius (r^2). Pi (π) is equal to 3.14.

Example
If your diameter is 10 cm,
$r = 5$ cm, $r^2 = 25$ cm^2, $\pi = 3.14$

$A = \pi r^2$
$A = 3.14 \times 25$ cm^2
$A = 78.5$ cm^2

6. Record your answers in your ScienceLog.

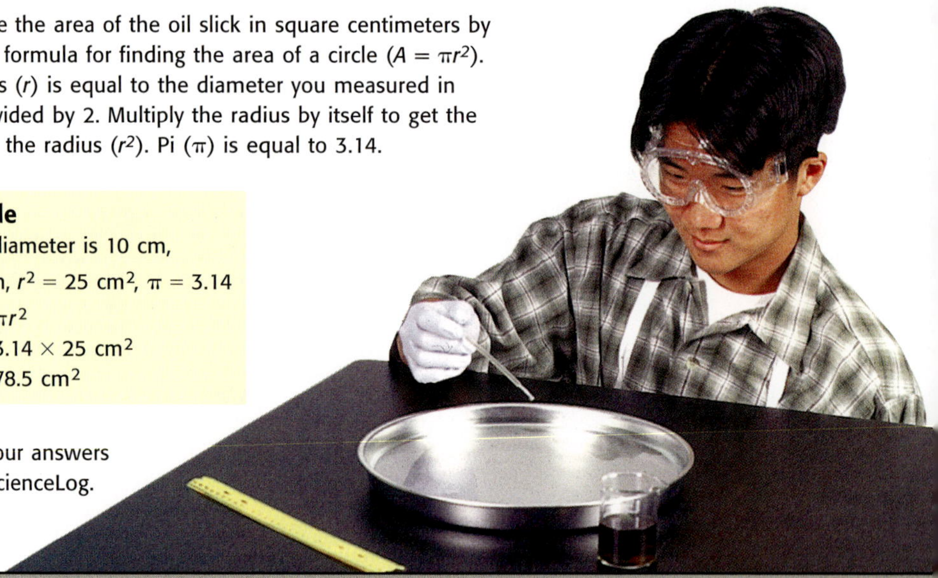

104

CLASSROOM TESTED & APPROVED

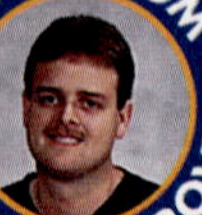

David Sparks
Redwater Jr. High
Redwater, Texas

Analysis

7. What happened to the drop of oil when it came in contact with the water? Did this surprise you?
8. What total surface area was covered by the oil slick? (Be sure to show your calculations.)
9. What does this experiment tell you about the density of oil compared with the density of water? Explain.

Going Further

Can you devise a way to clean the oil from the water? Get permission from your teacher before testing your cleaning method.

Do you think oil behaves the same way in ocean water? Devise an experiment to test your hypothesis.

Finding the Number of Drops in a Liter

"It's only a few drops," you may think as you spill something toxic on the ground. But those drops eventually add up. Just how many drops does it take to make a difference? In this activity, you'll learn just what an impact a few drops can have.

Procedure

10. Using a clean pipet, count the number of water drops it takes to fill the graduated cylinder to 10 mL. Be sure to add the drops slowly so you get an accurate count.
11. Since there are 1,000 mL in a liter, multiply the number of drops in 10 mL by 100. This gives you the number of drops in a liter.

Analysis

12. How many drops of water from your pipet does it take to fill a 1 L container?
13. What would happen if someone spilled 4 L of oil into a lake?

Going Further

Find out how much oil supertankers contain. Can you imagine the size of an oil slick that would form if one of these tankers spilled its oil?

LabBook

Answers

7. When the drop of oil touched the water, it spread out. This may surprise some students, who might expect the two substances to mix or the oil to sink.
8. Answers will vary. Students should show their work to illustrate that they understand the mathematical principles involved.
9. This experiment shows that oil is less dense than water. For this reason, oil floats on water.

Going Further

- Answers will vary. One possibility is to use detergent to change the surface tension and remove the oil.
- To test the behavior of oil in ocean water, students could repeat the activity, using cold salt water instead of tap water. They could also rock the pan back and forth to simulate waves.

12. Answers will vary. The answer depends on the number of drops the students count in 10 mL of water. They should show their work to illustrate that they understand the mathematical principles involved.
13. The oil would spread to cover a large area. Students will not be able to determine the exact area for such an oil slick, but they should understand that the oil will spread significantly and pollute much of the lake.

Datasheets for LabBook

Turning the Tides
Teacher's Notes

Time Required
One 45-minute class period

Lab Ratings
EASY → HARD

Teacher Prep: 2
Student Set-Up: 3
Concept Level: 3
Clean Up: 1

MATERIALS

The materials listed on the student page are enough for a group of 2–4 students.

Safety Caution
Remind students to review all safety cautions and icons before beginning this lab activity. Students should wear safety goggles. Be sure that students have enough space to spin the system.

Preparation Notes
You will need to put a mark at the center of all of the cardboard disks. Students will need to draw the radius of the circle from the center to the edge of the disk. Encourage students to draw this line along the corrugations of the cardboard. Otherwise, several other steps will be made more difficult. The cardboard disks are not to scale with the Earth and moon. They are used to show how a two-body system, such as the Earth-moon system, rotates. The disks must be different sizes. The large disks could be 10 cm in diameter, and the smaller disks could be 5 cm in diameter.

Turning the Tides

MAKING MODELS

Daily tides are caused by two "bulges" on the ocean's surface—one on the side of the Earth facing the moon and the other on the opposite side. The bulge on the side facing the moon is caused by the moon's gravitational pull on the water. But the bulge on the opposite side is slightly more difficult to explain. Whereas the moon pulls the water on one side of the Earth, the combined rotation of the Earth and the moon "pushes" the water on the opposite side of the Earth. In this activity, you will model the motion of the Earth and the moon to investigate the tidal bulge on the side of Earth facing away from the moon.

Materials
- 2 disks of corrugated cardboard, one large and one small, with centers marked
- white glue
- piece of dowel, 1/4 in. in diameter and 36 cm long
- 5 cm length of string
- stapler with staples
- 1 × 1 cm piece of cardboard
- sharp pencil

Procedure
1. Draw a line from the center of each disk along the folds in the cardboard to the edge of the disk. This line is the radius.
2. Place a drop of white glue on one end of the dowel. Lay the larger disk flat, and align the dowel with the line for the radius you drew in step 1. Insert about 2.5 cm of the dowel into the edge of the disk.
3. Add a drop of glue to the other end of the dowel, and push that end into the smaller disk, again along its radius. The setup should look like a large two-headed lollipop, as shown below. This is a model of the Earth-moon system.
4. Staple the string to the edge of the large disk on the side opposite the dowel. Staple the cardboard square to the other end of the string. This smaller piece of cardboard represents the Earth's oceans that face away from the moon.

5. Place the tip of the pencil at the center of the large disk, as shown in the figure on the next page, and spin the model. You may poke a small hole in the bottom of the disk with your pencil, but DO NOT poke all the way through the cardboard. Record your observations in your ScienceLog. **Caution:** Be sure you are at a safe distance from other people before spinning your model.

106

CLASSROOM TESTED & APPROVED

Tracy Jahn
Berkshire Jr.-Sr. High
Canaan, New York

6. Now find your model's *center of mass.* This is the point at which the model can be balanced on the end of the pencil. **Hint:** It might be easier to find the center of mass using the eraser end. Then use the sharpened end of the pencil to balance the model. This balance point should be just inside the edge of the larger disk.

7. Place the pencil at the center of mass, and spin the model around the pencil. Again, you may wish to poke a small hole in the disk. Record your observations in your ScienceLog.

Analysis

8. What happened when you tried to spin the model around the center of the large disk? This model, called the Earth-centered model, represents the incorrect view that the moon orbits the center of the Earth.

9. What happened when you tried to spin the model around its center of mass? This point, called the *barycenter,* is the point around which both the Earth and the moon rotate.

10. In each case, what happened to the string and cardboard square when the model was spun?

11. Which model—the Earth-centered model or the barycentric model—explains why the Earth has a tidal bulge on the side opposite the moon? Explain.

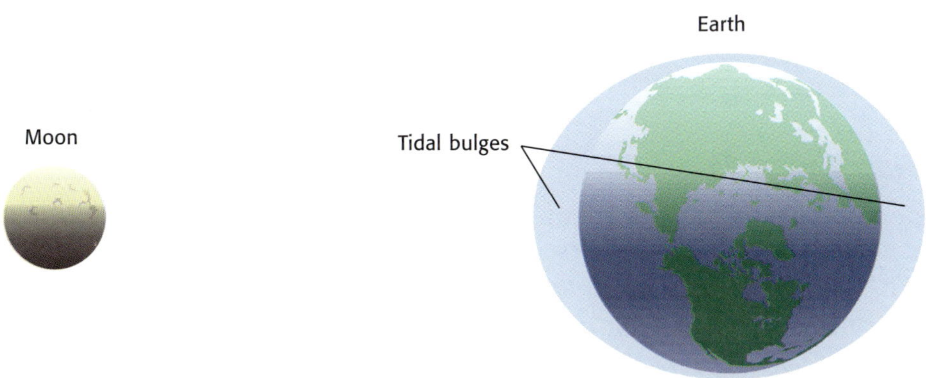

LabBook

Answers

8. Answers will vary. Sample answer: When I tried to spin the model around the center of the large disk, I could not get the model to balance on the pencil.

9. Answers will vary. Sample answer: I was able to balance the model at the barycenter. The model spun, and the small piece of cardboard on the string swung outward.

10. Answers will vary. Sample answer: The cardboard square hung down when I tried to swing the model around the center of the large disk; I was unable to make the model spin. When I spun the model around its barycenter, the square swung away from the model.

11. The barycentric model explains why the Earth has a bulge on the side opposite the moon. The side of the Earth opposite the side facing the moon acts in much the same way the small square of cardboard does in this model. As the Earth-moon system rotates, the side of the Earth facing away from the moon bulges outward.

Datasheets for LabBook

Concept Mapping: A Way to Bring Ideas Together

What Is a Concept Map?

Have you ever tried to tell someone about a book or a chapter you've just read and found that you can remember only a few isolated words and ideas? Or maybe you've memorized facts for a test and then weeks later discovered you're not even sure what topics those facts covered.

In both cases, you may have understood the ideas or concepts by themselves but not in relation to one another. If you could somehow link the ideas together, you would probably understand them better and remember them longer. This is something a concept map can help you do. A concept map is a way to see how ideas or concepts fit together. It can help you see the "big picture."

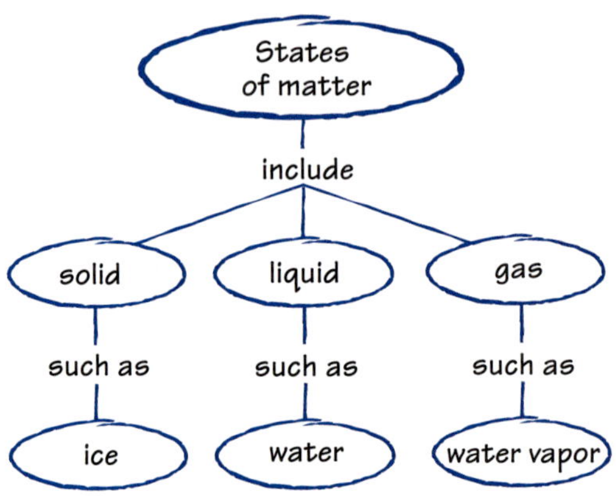

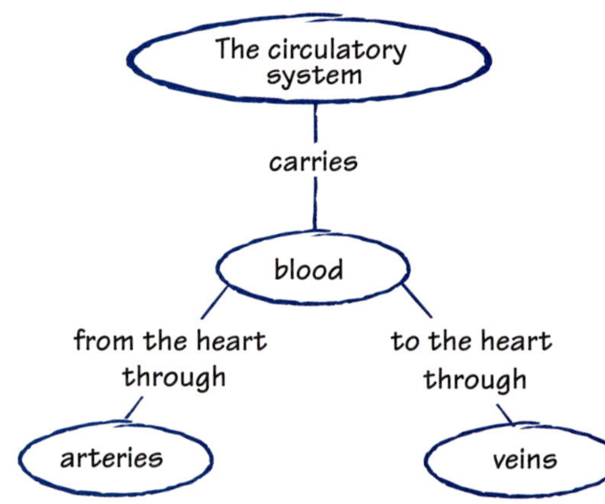

How to Make a Concept Map

1. **Make a list of the main ideas or concepts.**

   It might help to write each concept on its own slip of paper. This will make it easier to rearrange the concepts as many times as necessary to make sense of how the concepts are connected. After you've made a few concept maps this way, you can go directly from writing your list to actually making the map.

2. **Arrange the concepts in order from the most general to the most specific.**

   Put the most general concept at the top and circle it. Ask yourself, "How does this concept relate to the remaining concepts?" As you see the relationships, arrange the concepts in order from general to specific.

3. **Connect the related concepts with lines.**

4. **On each line, write an action word or short phrase that shows how the concepts are related.**

Look at the concept maps on this page, and then see if you can make one for the following terms:

plants, water, photosynthesis, carbon dioxide, sun's energy

One possible answer is provided at right, but don't look at it until you try the concept map yourself.

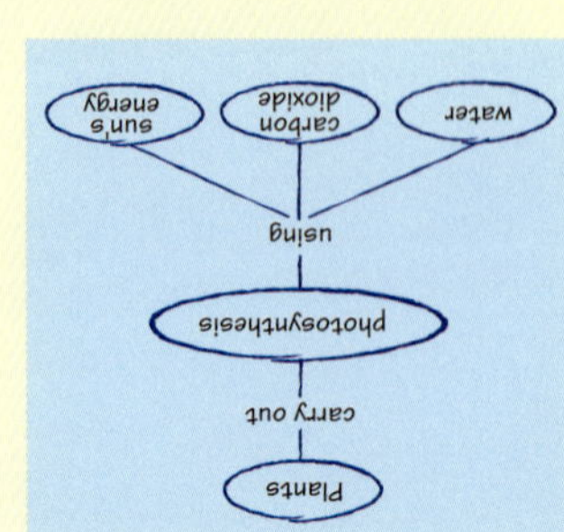

SI Measurement

The International System of Units, or SI, is the standard system of measurement used by many scientists. Using the same standards of measurement makes it easier for scientists to communicate with one another.

SI works by combining prefixes and base units. Each base unit can be used with different prefixes to define smaller and larger quantities. The table below lists common SI prefixes.

SI Prefixes

Prefix	Abbreviation	Factor	Example
kilo-	k	1,000	kilogram, 1 kg = 1,000 g
hecto-	h	100	hectoliter, 1 hL = 100 L
deka-	da	10	dekameter, 1 dam = 10 m
		1	meter, liter
deci-	d	0.1	decigram, 1 dg = 0.1 g
centi-	c	0.01	centimeter, 1 cm = 0.01 m
milli-	m	0.001	milliliter, 1 mL = 0.001 L
micro-	μ	0.000 001	micrometer, 1 μm = 0.000 001 m

SI Conversion Table

SI units	From SI to English	From English to SI
Length		
kilometer (km) = 1,000 m	1 km = 0.621 mi	1 mi = 1.609 km
meter (m) = 100 cm	1 m = 3.281 ft	1 ft = 0.305 m
centimeter (cm) = 0.01 m	1 cm = 0.394 in.	1 in. = 2.540 cm
millimeter (mm) = 0.001 m	1 mm = 0.039 in.	
micrometer (μm) = 0.000 001 m		
nanometer (nm) = 0.000 000 001 m		
Area		
square kilometer (km^2) = 100 hectares	1 km^2 = 0.386 mi^2	1 mi^2 = 2.590 km^2
hectare (ha) = 10,000 m^2	1 ha = 2.471 acres	1 acre = 0.405 ha
square meter (m^2) = 10,000 cm^2	1 m^2 = 10.765 ft^2	1 ft^2 = 0.093 m^2
square centimeter (cm^2) = 100 mm^2	1 cm^2 = 0.155 $in.^2$	1 $in.^2$ = 6.452 cm^2
Volume		
liter (L) = 1,000 mL = 1 dm^3	1 L = 1.057 fl qt	1 fl qt = 0.946 L
milliliter (mL) = 0.001 L = 1 cm^3	1 mL = 0.034 fl oz	1 fl oz = 29.575 mL
microliter (μL) = 0.000 001 L		
Mass		
kilogram (kg) = 1,000 g	1 kg = 2.205 lb	1 lb = 0.454 kg
gram (g) = 1,000 mg	1 g = 0.035 oz	1 oz = 28.349 g
milligram (mg) = 0.001 g		
microgram (μg) = 0.000 001 g		

Temperature Scales

Temperature can be expressed using three different scales: Fahrenheit, Celsius, and Kelvin. The SI unit for temperature is the kelvin (K). Although 0 K is much colder than 0°C, a change of 1 K is equal to a change of 1°C.

Three Temperature Scales

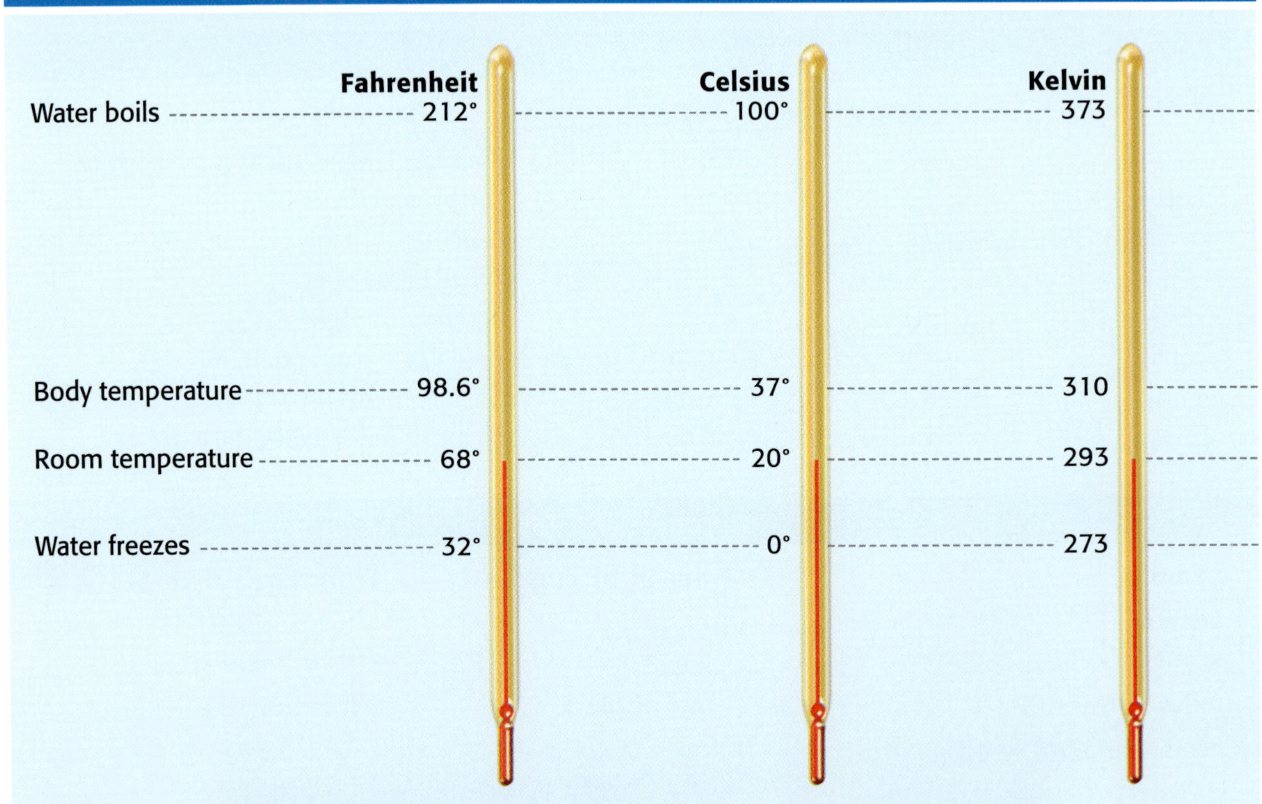

Temperature Conversions Table

To convert	Use this equation:	Example
Celsius to Fahrenheit °C ⟶ °F	$°F = \left(\frac{9}{5} \times °C\right) + 32$	Convert 45°C to °F. $°F = \left(\frac{9}{5} \times 45°C\right) + 32 = 113°F$
Fahrenheit to Celsius °F ⟶ °C	$°C = \frac{5}{9} \times (°F - 32)$	Convert 68°F to °C. $°C = \frac{5}{9} \times (68°F - 32) = 20°C$
Celsius to Kelvin °C ⟶ K	$K = °C + 273$	Convert 45°C to K. $K = 45°C + 273 = 318\ K$
Kelvin to Celsius K ⟶ °C	$°C = K - 273$	Convert 32 K to °C. $°C = 32\ K - 273 = -241°C$

Measuring Skills

Using a Graduated Cylinder

When using a graduated cylinder to measure volume, keep the following procedures in mind:

1. Make sure the cylinder is on a flat, level surface.
2. Move your head so that your eye is level with the surface of the liquid.
3. Read the mark closest to the liquid level. On glass graduated cylinders, read the mark closest to the center of the curve in the liquid's surface.

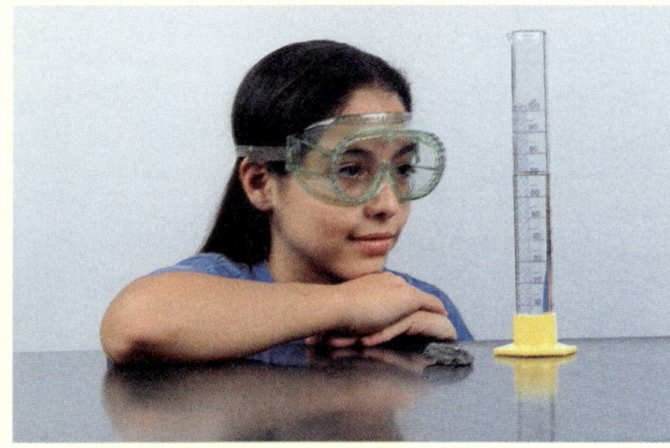

Using a Meterstick or Metric Ruler

When using a meterstick or metric ruler to measure length, keep the following procedures in mind:

1. Place the ruler firmly against the object you are measuring.
2. Align one edge of the object exactly with the zero end of the ruler.
3. Look at the other edge of the object to see which of the marks on the ruler is closest to that edge. **Note:** Each small slash between the centimeters represents a millimeter, which is one-tenth of a centimeter.

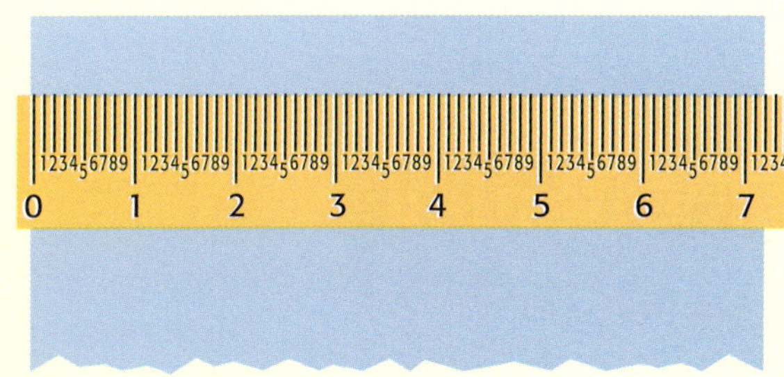

Using a Triple-Beam Balance

When using a triple-beam balance to measure mass, keep the following procedures in mind:

1. Make sure the balance is on a level surface.
2. Place all of the countermasses at zero. Adjust the balancing knob until the pointer rests at zero.
3. Place the object you wish to measure on the pan. **Caution:** Do not place hot objects or chemicals directly on the balance pan.
4. Move the largest countermass along the beam to the right until it is at the last notch that does not tip the balance. Follow the same procedure with the next-largest countermass. Then move the smallest countermass until the pointer rests at zero.
5. Add the readings from the three beams together to determine the mass of the object.
6. When determining the mass of crystals or powders, use a piece of filter paper. First find the mass of the paper. Then add the crystals or powder to the paper and re-measure. The actual mass of the crystals or powder is the total mass minus the mass of the paper. When finding the mass of liquids, first find the mass of the empty container. Then find the mass of the liquid and container together. The mass of the liquid is the total mass minus the mass of the container.

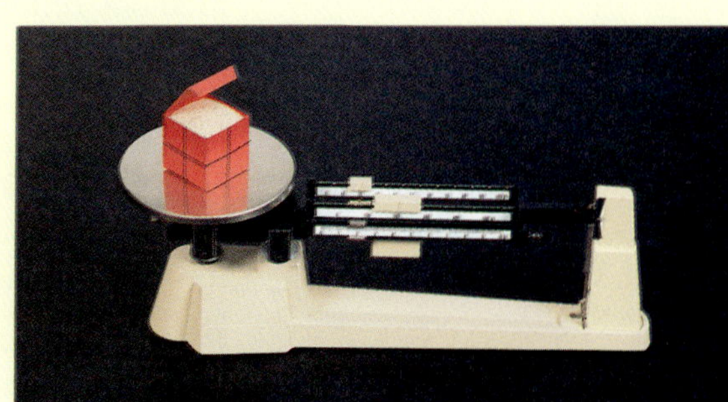

Scientific Method

The series of steps that scientists use to answer questions and solve problems is often called the **scientific method.** The scientific method is not a rigid procedure. Scientists may use all of the steps or just some of the steps of the scientific method. They may even repeat some of the steps. The goal of the scientific method is to come up with reliable answers and solutions.

Six Steps of the Scientific Method

1 **Ask a Question** Good questions come from careful **observations.** You make observations by using your senses to gather information. Sometimes you may use instruments, such as microscopes and telescopes, to extend the range of your senses. As you observe the natural world, you will discover that you have many more questions than answers. These questions drive the scientific method.

Questions beginning with *what, why, how,* and *when* are very important in focusing an investigation, and they often lead to a hypothesis. (You will learn what a hypothesis is in the next step.) Here is an example of a question that could lead to further investigation.

Question: How does acid rain affect plant growth?

2 **Form a Hypothesis** After you come up with a question, you need to turn the question into a **hypothesis.** A hypothesis is a clear statement of what you expect the answer to your question to be. Your hypothesis will represent your best "educated guess" based on your observations and what you already know. A good hypothesis is testable. If observations and information cannot be gathered or if an experiment cannot be designed to test your hypothesis, it is untestable, and the investigation can go no further.

Here is a hypothesis that could be formed from the question, "How does acid rain affect plant growth?"

Hypothesis: Acid rain causes plants to grow more slowly.

Notice that the hypothesis provides some specifics that lead to methods of testing. The hypothesis can also lead to predictions. A **prediction** is what you think will be the outcome of your experiment or data collection. Predictions are usually stated in an "if . . . then" format. For example, **if** meat is kept at room temperature, **then** it will spoil faster than meat kept in the refrigerator. More than one prediction can be made for a single hypothesis. Here is a sample prediction for the hypothesis that acid rain causes plants to grow more slowly.

Prediction: If a plant is watered with only acid rain (which has a pH of 4), then the plant will grow at half its normal rate.

3 **Test the Hypothesis** After you have formed a hypothesis and made a prediction, you should test your hypothesis. There are different ways to do this. Perhaps the most familiar way is to conduct a **controlled experiment.** A controlled experiment tests only one factor at a time. A controlled experiment has a **control group** and one or more **experimental groups.** All the factors for the control and experimental groups are the same except for one factor, which is called the **variable.** By changing only one factor, you can see the results of just that one change.

Sometimes, the nature of an investigation makes a controlled experiment impossible. For example, dinosaurs have been extinct for millions of years, and the Earth's core is surrounded by thousands of meters of rock. It would be difficult, if not impossible, to conduct controlled experiments on such things. Under such circumstances, a hypothesis may be tested by making detailed observations. Taking measurements is one way of making observations.

4 **Analyze the Results** After you have completed your experiments, made your observations, and collected your data, you must analyze all the information you have gathered. Tables and graphs are often used in this step to organize the data.

5 **Draw Conclusions** Based on the analysis of your data, you should conclude whether or not your results support your hypothesis. If your hypothesis is supported, you (or others) might want to repeat the observations or experiments to verify your results. If your hypothesis is not supported by the data, you may have to check your procedure for errors. You may even have to reject your hypothesis and make a new one. If you cannot draw a conclusion from your results, you may have to try the investigation again or carry out further observations or experiments.

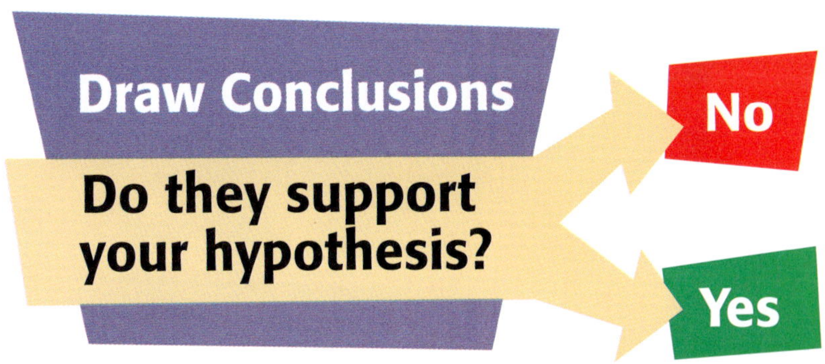

6 **Communicate Results** After any scientific investigation, you should report your results. By doing a written or oral report, you let others know what you have learned. They may want to repeat your investigation to see if they get the same results. Your report may even lead to another question, which in turn may lead to another investigation.

Scientific Method in Action

The scientific method is not a "straight line" of steps. It contains loops in which several steps may be repeated over and over again, while others may not be necessary. For example, sometimes scientists will find that testing one hypothesis raises new questions and new hypotheses to be tested. And sometimes, testing the hypothesis leads directly to a conclusion. Furthermore, the steps in the scientific method are not always used in the same order. Follow the steps in the diagram below, and see how many different directions the scientific method can take you.

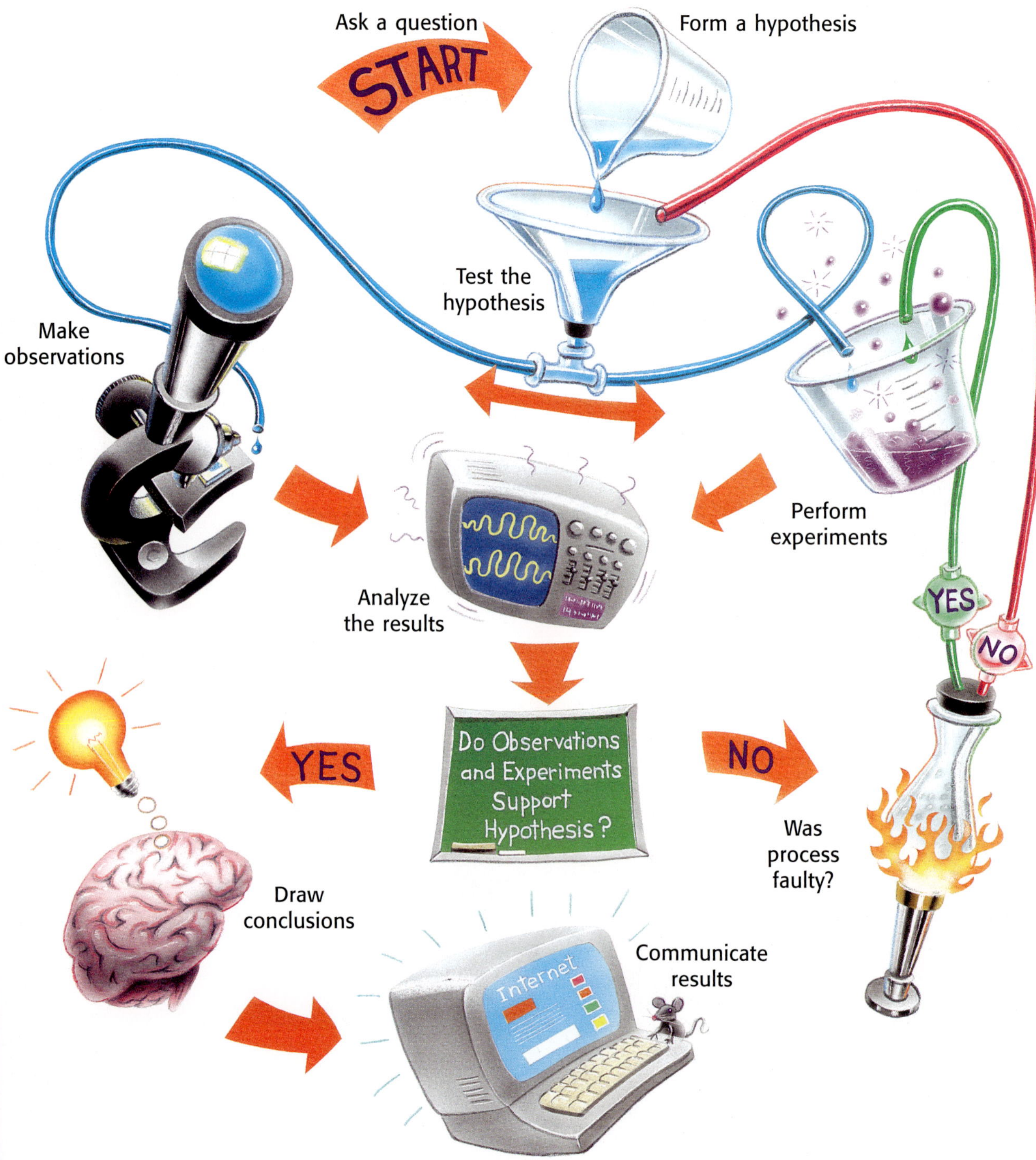

Making Charts and Graphs

Circle Graphs

A circle graph, or pie chart, shows how each group of data relates to all of the data. Each part of the circle represents a category of the data. The entire circle represents all of the data. For example, a biologist studying a hardwood forest in Wisconsin found that there were five different types of trees. The data table at right summarizes the biologist's findings.

Wisconsin Hardwood Trees	
Type of tree	**Number found**
Oak	600
Maple	750
Beech	300
Birch	1,200
Hickory	150
Total	3,000

How to Make a Circle Graph

1. In order to make a circle graph of this data, first find the percentage of each type of tree. To do this, divide the number of individual trees by the total number of trees and multiply by 100.

$$\frac{600 \text{ oak}}{3{,}000 \text{ trees}} \times 100 = 20\%$$

$$\frac{750 \text{ maple}}{3{,}000 \text{ trees}} \times 100 = 25\%$$

$$\frac{300 \text{ beech}}{3{,}000 \text{ trees}} \times 100 = 10\%$$

$$\frac{1{,}200 \text{ birch}}{3{,}000 \text{ trees}} \times 100 = 40\%$$

$$\frac{150 \text{ hickory}}{3{,}000 \text{ trees}} \times 100 = 5\%$$

2. Now determine the size of the pie shapes that make up the chart. Do this by multiplying each percentage by 360°. Remember that a circle contains 360°.

$20\% \times 360° = 72°$ $\quad$ $25\% \times 360° = 90°$
$10\% \times 360° = 36°$ $\quad$ $40\% \times 360° = 144°$
$5\% \times 360° = 18°$

3. Then check that the sum of the percentages is 100 and the sum of the degrees is 360.

$20\% + 25\% + 10\% + 40\% + 5\% = 100\%$

$72° + 90° + 36° + 144° + 18° = 360°$

4. Use a compass to draw a circle and mark its center.

5. Then use a protractor to draw angles of 72°, 90°, 36°, 144°, and 18° in the circle.

6. Finally, label each part of the graph, and choose an appropriate title.

A Community of Wisconsin Hardwood Trees

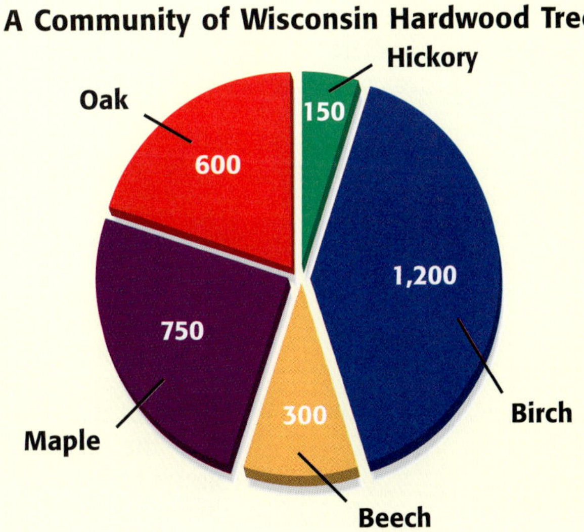

115

Line Graphs

Population of Appleton, 1900–2000	
Year	**Population**
1900	1,800
1920	2,500
1940	3,200
1960	3,900
1980	4,600
2000	5,300

Line graphs are most often used to demonstrate continuous change. For example, Mr. Smith's science class analyzed the population records for their hometown, Appleton, between 1900 and 2000. Examine the data at left.

Because the year and the population change, they are the *variables*. The population is determined by, or dependent on, the year. Therefore, the population is called the **dependent variable**, and the year is called the **independent variable.** Each set of data is called a **data pair.** To prepare a line graph, data pairs must first be organized in a table like the one at left.

How to Make a Line Graph

1. Place the independent variable along the horizontal (x) axis. Place the dependent variable along the vertical (y) axis.
2. Label the x-axis "Year" and the y-axis "Population." Look at your largest and smallest values for the population. Determine a scale for the y-axis that will provide enough space to show these values. You must use the same scale for the entire length of the axis. Find an appropriate scale for the x-axis too.
3. Choose reasonable starting points for each axis.
4. Plot the data pairs as accurately as possible.
5. Choose a title that accurately represents the data.

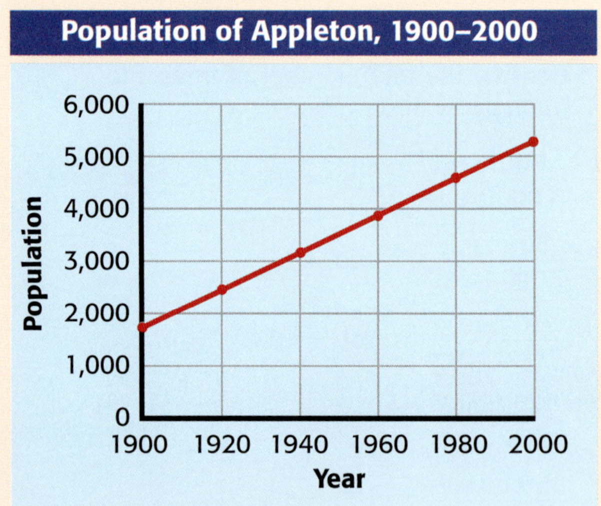

How to Determine Slope

Slope is the ratio of the change in the y-axis to the change in the x-axis, or "rise over run."

1. Choose two points on the line graph. For example, the population of Appleton in 2000 was 5,300 people. Therefore, you can define point a as (2000, 5,300). In 1900, the population was 1,800 people. Define point b as (1900, 1,800).
2. Find the change in the y-axis.
   (y at point a) − (y at point b)
   5,300 people − 1,800 people = 3,500 people
3. Find the change in the x-axis.
   (x at point a) − (x at point b)
   2000 − 1900 = 100 years
4. Calculate the slope of the graph by dividing the change in y by the change in x.

$$\text{slope} = \frac{\text{change in } y}{\text{change in } x}$$

$$\text{slope} = \frac{3{,}500 \text{ people}}{100 \text{ years}}$$

$$\text{slope} = 35 \text{ people per year}$$

In this example, the population in Appleton increased by a fixed amount each year. The graph of this data is a straight line. Therefore, the relationship is **linear.** When the graph of a set of data is not a straight line, the relationship is **nonlinear.**

116

Using Algebra to Determine Slope

The equation in step 4 may also be arranged to be:

$$y = kx$$

where y represents the change in the y-axis, k represents the slope, and x represents the change in the x-axis.

$$\text{slope} = \frac{\text{change in } y}{\text{change in } x}$$

$$k = \frac{y}{x}$$

$$k \times x = \frac{y \times x}{x}$$

$$kx = y$$

Bar Graphs

Bar graphs are used to demonstrate change that is not continuous. These graphs can be used to indicate trends when the data are taken over a long period of time. A meteorologist gathered the precipitation records at right for Hartford, Connecticut, for April 1–15, 1996, and used a bar graph to represent the data.

Precipitation in Hartford, Connecticut April 1–15, 1996

Date	Precipitation (cm)	Date	Precipitation (cm)
April 1	0.5	April 9	0.25
April 2	1.25	April 10	0.0
April 3	0.0	April 11	1.0
April 4	0.0	April 12	0.0
April 5	0.0	April 13	0.25
April 6	0.0	April 14	0.0
April 7	0.0	April 15	6.50
April 8	1.75		

How to Make a Bar Graph

1. Use an appropriate scale and a reasonable starting point for each axis.
2. Label the axes, and plot the data.
3. Choose a title that accurately represents the data.

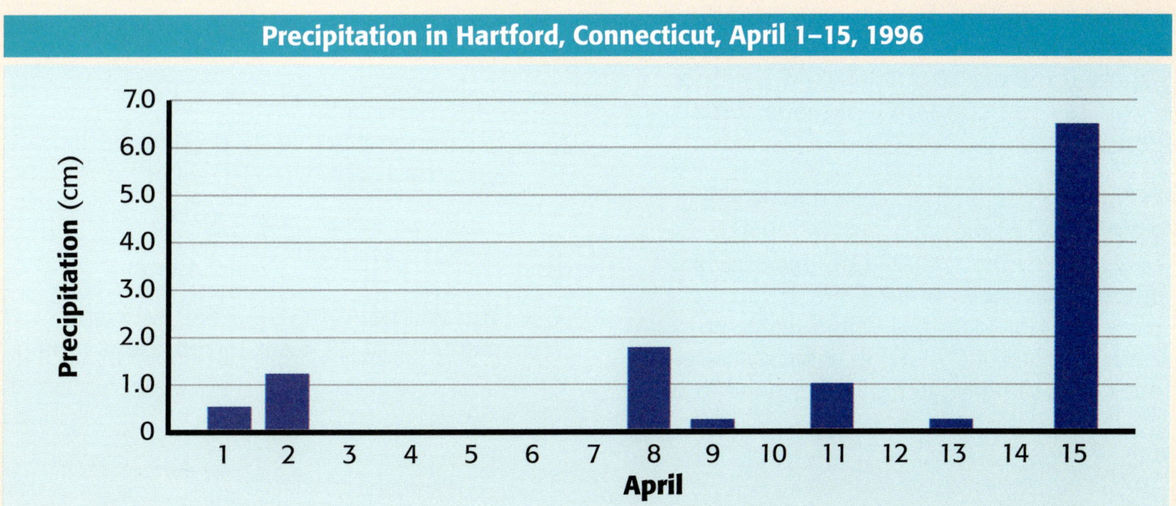

Appendix

Math Refresher

Science requires an understanding of many math concepts. The following pages will help you review some important math skills.

Averages

An **average,** or **mean,** simplifies a list of numbers into a single number that *approximates* their value.

Example: Find the average of the following set of numbers: 5, 4, 7, and 8.

Step 1: Find the sum.

$$5 + 4 + 7 + 8 = 24$$

Step 2: Divide the sum by the amount of numbers in your set. Because there are four numbers in this example, divide the sum by 4.

$$\frac{24}{4} = 6$$

The average, or mean, is **6.**

Ratios

A **ratio** is a comparison between numbers, and it is usually written as a fraction.

Example: Find the ratio of thermometers to students if you have 36 thermometers and 48 students in your class.

Step 1: Make the ratio.

$$\frac{36 \text{ thermometers}}{48 \text{ students}}$$

Step 2: Reduce the fraction to its simplest form.

$$\frac{36}{48} = \frac{36 \div 12}{48 \div 12} = \frac{3}{4}$$

The ratio of thermometers to students is **3 to 4,** or $\frac{3}{4}$. The ratio may also be written in the form 3:4.

Proportions

A **proportion** is an equation that states that two ratios are equal.

$$\frac{3}{1} = \frac{12}{4}$$

To solve a proportion, first multiply across the equal sign. This is called cross-multiplication. If you know three of the quantities in a proportion, you can use cross-multiplication to find the fourth.

Example: Imagine that you are making a scale model of the solar system for your science project. The diameter of Jupiter is 11.2 times the diameter of the Earth. If you are using a plastic-foam ball with a diameter of 2 cm to represent the Earth, what diameter does the ball representing Jupiter need to be?

$$\frac{11.2}{1} = \frac{x}{2 \text{ cm}}$$

Step 1: Cross-multiply.

$$\frac{11.2}{1} \times \frac{x}{2}$$

$$11.2 \times 2 = x \times 1$$

Step 2: Multiply.

$$22.4 = x \times 1$$

Step 3: Isolate the variable by dividing both sides by 1.

$$x = \frac{22.4}{1}$$

$$x = 22.4 \text{ cm}$$

You will need to use a ball with a diameter of **22.4 cm** to represent Jupiter.

Percentages

A **percentage** is a ratio of a given number to 100.

Example: What is 85 percent of 40?

Step 1: Rewrite the percentage by moving the decimal point two places to the left.

$$.85$$

Step 2: Multiply the decimal by the number you are calculating the percentage of.

$$0.85 \times 40 = 34$$

85 percent of 40 is **34.**

Decimals

To **add** or **subtract decimals,** line up the digits vertically so that the decimal points line up. Then add or subtract the columns from right to left, carrying or borrowing numbers as necessary.

Example: Add the following numbers: 3.1415 and 2.96.

Step 1: Line up the digits vertically so that the decimal points line up.

$$\begin{array}{r} 3.1415 \\ +\ 2.96 \\ \hline \end{array}$$

Step 2: Add the columns from right to left, carrying when necessary.

$$\begin{array}{r} {\scriptstyle 1\ 1} \\ 3.1415 \\ +\ 2.96 \\ \hline 6.1015 \end{array}$$

The sum is **6.1015.**

Fractions

Numbers tell you how many; **fractions** tell you *how much of a whole.*

Example: Your class has 24 plants. Your teacher instructs you to put 5 in a shady spot. What fraction does this represent?

Step 1: Write a fraction with the total number of parts in the whole as the denominator.

$$\frac{?}{24}$$

Step 2: Write the number of parts of the whole being represented as the numerator.

$$\frac{5}{24}$$

$\mathbf{\frac{5}{24}}$ of the plants will be in the shade.

Reducing Fractions

It is usually best to express a fraction in simplest form. This is called *reducing* a fraction.

Example: Reduce the fraction $\frac{30}{45}$ to its simplest form.

Step 1: Find the largest whole number that will divide evenly into both the numerator and denominator. This number is called the greatest common factor (GCF).

factors of the numerator 30: 1, 2, 3, 5, 6, 10, **15,** 30

factors of the denominator 45: 1, 3, 5, 9, **15,** 45

Step 2: Divide both the numerator and the denominator by the GCF, which in this case is 15.

$$\frac{30}{45} = \frac{30 \div 15}{45 \div 15} = \frac{2}{3}$$

$\frac{30}{45}$ reduced to its simplest form is $\mathbf{\frac{2}{3}}$.

119

Adding and Subtracting Fractions

To **add** or **subtract fractions** that have the **same denominator,** simply add or subtract the numerators.

Examples:

$$\frac{3}{5} + \frac{1}{5} = ? \text{ and } \frac{3}{4} - \frac{1}{4} = ?$$

Step 1: Add or subtract the numerators.

$$\frac{3}{5} + \frac{1}{5} = \frac{4}{\ } \text{ and } \frac{3}{4} - \frac{1}{4} = \frac{2}{\ }$$

Step 2: Write the sum or difference over the denominator.

$$\frac{3}{5} + \frac{1}{5} = \frac{4}{5} \text{ and } \frac{3}{4} - \frac{1}{4} = \frac{2}{4}$$

Step 3: If necessary, reduce the fraction to its simplest form.

$\mathbf{\frac{4}{5}}$ cannot be reduced, and $\frac{2}{4} = \mathbf{\frac{1}{2}}$.

To **add** or **subtract fractions** that have **different denominators,** first find the least common denominator (LCD).

Examples:

$$\frac{1}{2} + \frac{1}{6} = ? \text{ and } \frac{3}{4} - \frac{2}{3} = ?$$

Step 1: Write the equivalent fractions with a common denominator.

$$\frac{3}{6} + \frac{1}{6} = ? \text{ and } \frac{9}{12} - \frac{8}{12} = ?$$

Step 2: Add or subtract.

$$\frac{3}{6} + \frac{1}{6} = \frac{4}{6} \text{ and } \frac{9}{12} - \frac{8}{12} = \frac{1}{12}$$

Step 3: If necessary, reduce the fraction to its simplest form.

$\frac{4}{6} = \mathbf{\frac{2}{3}}$, and $\mathbf{\frac{1}{12}}$ cannot be reduced.

Multiplying Fractions

To **multiply fractions,** multiply the numerators and the denominators together, and then reduce the fraction to its simplest form.

Example:

$$\frac{5}{9} \times \frac{7}{10} = ?$$

Step 1: Multiply the numerators and denominators.

$$\frac{5}{9} \times \frac{7}{10} = \frac{5 \times 7}{9 \times 10} = \frac{35}{90}$$

Step 2: Reduce.

$$\frac{35}{90} = \frac{35 \div 5}{90 \div 5} = \mathbf{\frac{7}{18}}$$

Dividing Fractions

To **divide fractions**, first rewrite the divisor (the number you divide *by*) upside down. This is called the reciprocal of the divisor. Then you can multiply and reduce if necessary.

Example:

$$\frac{5}{8} \div \frac{3}{2} = ?$$

Step 1: Rewrite the divisor as its reciprocal.

$$\frac{3}{2} \rightarrow \frac{2}{3}$$

Step 2: Multiply.

$$\frac{5}{8} \times \frac{2}{3} = \frac{5 \times 2}{8 \times 3} = \frac{10}{24}$$

Step 3: Reduce.

$$\frac{10}{24} = \frac{10 \div 2}{24 \div 2} = \mathbf{\frac{5}{12}}$$

Scientific Notation

Scientific notation is a short way of representing very large and very small numbers without writing all of the place-holding zeros.

Example: Write 653,000,000 in scientific notation.

Step 1: Write the number without the place-holding zeros.

653

Step 2: Place the decimal point after the first digit.

6.53

Step 3: Find the exponent by counting the number of places that you moved the decimal point.

6.53000000

The decimal point was moved eight places to the left. Therefore, the exponent of 10 is positive 8. Remember, if the decimal point had moved to the right, the exponent would be negative.

Step 4: Write the number in scientific notation.

$$\mathbf{6.53 \times 10^8}$$

Area

Area is the number of square units needed to cover the surface of an object.

Formulas:

Area of a square = side × side

Area of a rectangle = length × width

Area of a triangle = $\frac{1}{2}$ × base × height

Examples: Find the areas.

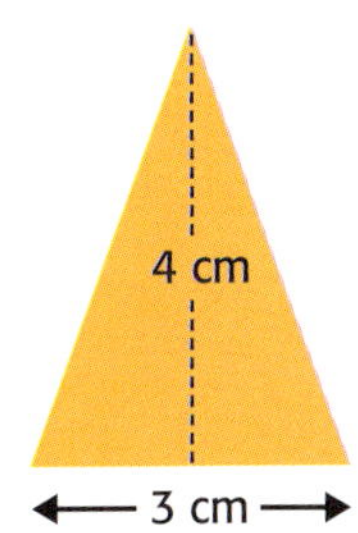

Triangle

Area = $\frac{1}{2}$ × base × height

Area = $\frac{1}{2}$ × 3 cm × 4 cm

Area = **6 cm²**

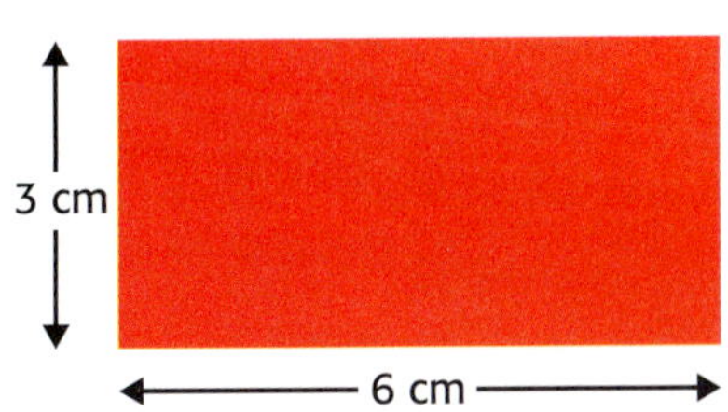

Rectangle

Area = length × width

Area = 6 cm × 3 cm

Area = $\mathbf{18\ cm^2}$

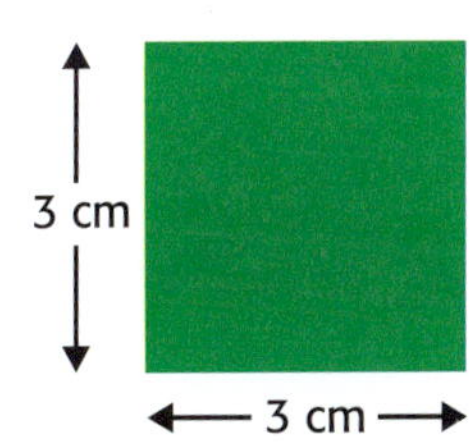

Square

Area = side × side

Area = 3 cm × 3 cm

Area = $\mathbf{9\ cm^2}$

Volume

Volume is the amount of space something occupies.

Formulas:

Volume of a cube = side × side × side

Volume of a prism = area of base × height

Examples:

Find the volume of the solids.

Cube

Volume = side × side × side

Volume = 4 cm × 4 cm × 4 cm

Volume = $\mathbf{64\ cm^3}$

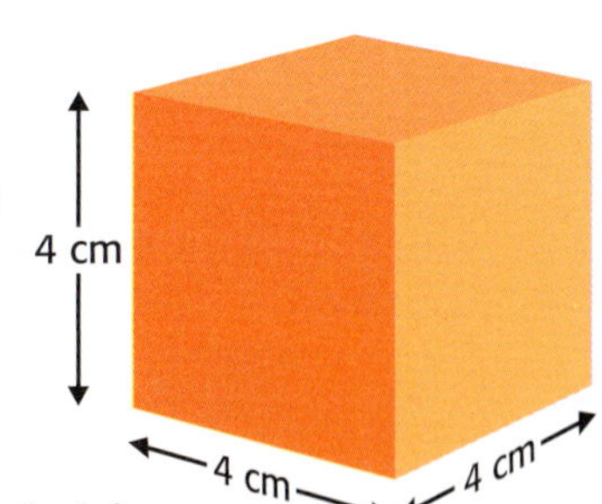

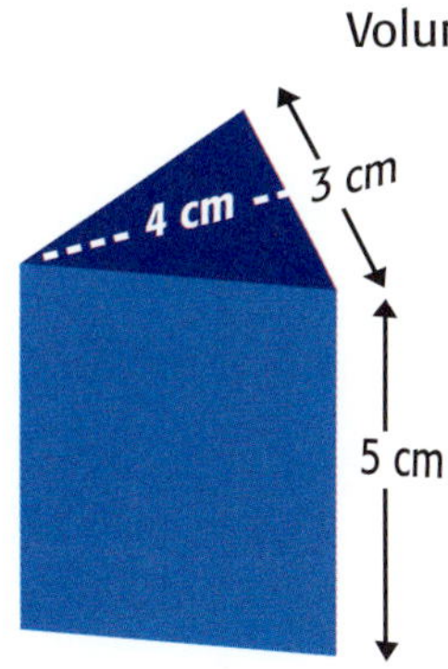

Prism

Volume = area of base × height

Volume = (area of triangle) × height

Volume = $\left(\frac{1}{2} \times 3\text{ cm} \times 4\text{ cm}\right) \times 5\text{ cm}$

Volume = $6\text{ cm}^2 \times 5\text{ cm}$

Volume = $\mathbf{30\ cm^3}$

Periodic Table of the Elements

Each square on the table includes an element's name, chemical symbol, atomic number, and atomic mass.

Atomic number — 6
Chemical symbol — C
Element name — Carbon
Atomic mass — 12.0

The background color indicates the type of element. Carbon is a nonmetal.

The color of the chemical symbol indicates the physical state at room temperature. Carbon is a solid.

Background
- Metals
- Metalloids
- Nonmetals

Chemical Symbol
- Solid
- Liquid
- Gas

	Group 1	Group 2	Group 3	Group 4	Group 5	Group 6	Group 7	Group 8	Group 9
Period 1	1 H Hydrogen 1.0								
Period 2	3 Li Lithium 6.9	4 Be Beryllium 9.0							
Period 3	11 Na Sodium 23.0	12 Mg Magnesium 24.3							
Period 4	19 K Potassium 39.1	20 Ca Calcium 40.1	21 Sc Scandium 45.0	22 Ti Titanium 47.9	23 V Vanadium 50.9	24 Cr Chromium 52.0	25 Mn Manganese 54.9	26 Fe Iron 55.8	27 Co Cobalt 58.9
Period 5	37 Rb Rubidium 85.5	38 Sr Strontium 87.6	39 Y Yttrium 88.9	40 Zr Zirconium 91.2	41 Nb Niobium 92.9	42 Mo Molybdenum 95.9	43 Tc Technetium (97.9)	44 Ru Ruthenium 101.1	45 Rh Rhodium 102.9
Period 6	55 Cs Cesium 132.9	56 Ba Barium 137.3	57 La Lanthanum 138.9	72 Hf Hafnium 178.5	73 Ta Tantalum 180.9	74 W Tungsten 183.8	75 Re Rhenium 186.2	76 Os Osmium 190.2	77 Ir Iridium 192.2
Period 7	87 Fr Francium (223.0)	88 Ra Radium (226.0)	89 Ac Actinium (227.0)	104 Rf Rutherfordium (261.1)	105 Db Dubnium (262.1)	106 Sg Seaborgium (263.1)	107 Bh Bohrium (262.1)	108 Hs Hassium (265)	109 Mt Meitnerium (266)

A row of elements is called a period.

A column of elements is called a group or family.

Lanthanides	58 Ce Cerium 140.1	59 Pr Praseodymium 140.9	60 Nd Neodymium 144.2	61 Pm Promethium (144.9)	62 Sm Samarium 150.4
Actinides	90 Th Thorium 232.0	91 Pa Protactinium 231.0	92 U Uranium 238.0	93 Np Neptunium (237.0)	94 Pu Plutonium 244.1

These elements are placed below the table to allow the table to be narrower.

122

This zigzag line reminds you where the metals, nonmetals, and metalloids are.

Group 10	Group 11	Group 12	Group 13	Group 14	Group 15	Group 16	Group 17	Group 18
								2 **He** Helium 4.0
			5 **B** Boron 10.8	6 **C** Carbon 12.0	7 **N** Nitrogen 14.0	8 **O** Oxygen 16.0	9 **F** Fluorine 19.0	10 **Ne** Neon 20.2
			13 **Al** Aluminum 27.0	14 **Si** Silicon 28.1	15 **P** Phosphorus 31.0	16 **S** Sulfur 32.1	17 **Cl** Chlorine 35.5	18 **Ar** Argon 39.9
28 **Ni** Nickel 58.7	29 **Cu** Copper 63.5	30 **Zn** Zinc 65.4	31 **Ga** Gallium 69.7	32 **Ge** Germanium 72.6	33 **As** Arsenic 74.9	34 **Se** Selenium 79.0	35 **Br** Bromine 79.9	36 **Kr** Krypton 83.8
46 **Pd** Palladium 106.4	47 **Ag** Silver 107.9	48 **Cd** Cadmium 112.4	49 **In** Indium 114.8	50 **Sn** Tin 118.7	51 **Sb** Antimony 121.8	52 **Te** Tellurium 127.6	53 **I** Iodine 126.9	54 **Xe** Xenon 131.3
78 **Pt** Platinum 195.1	79 **Au** Gold 197.0	80 **Hg** Mercury 200.6	81 **Tl** Thallium 204.4	82 **Pb** Lead 207.2	83 **Bi** Bismuth 209.0	84 **Po** Polonium (209.0)	85 **At** Astatine (210.0)	86 **Rn** Radon (222.0)
110 **Uun*** Ununnilium (271)	111 **Uuu*** Unununium (272)	112 **Uub*** Ununbium (277)		114 **Uuq*** Ununquadium (285)		116 **Uuh*** Ununhexium (289)		118 **Uuo*** Ununoctium (293)

A number in parenthesis is the mass number of the most stable form of that element.

63 **Eu** Europium 152.0	64 **Gd** Gadolinium 157.3	65 **Tb** Terbium 158.9	66 **Dy** Dysprosium 162.5	67 **Ho** Holmium 164.9	68 **Er** Erbium 167.3	69 **Tm** Thulium 168.9	70 **Yb** Ytterbium 173.0	71 **Lu** Lutetium 175.0
95 **Am** Americium (243.1)	96 **Cm** Curium (247.1)	97 **Bk** Berkelium (247.1)	98 **Cf** Californium (251.1)	99 **Es** Einsteinium (252.1)	100 **Fm** Fermium (257.1)	101 **Md** Mendelevium (258.1)	102 **No** Nobelium (259.1)	103 **Lr** Lawrencium (262.1)

**The official names and symbols for the elements greater than 109 will eventually be approved by a committee of scientists.*

Physical Laws and Equations

Law of Conservation of Energy

The law of conservation of energy states that energy can be neither created nor destroyed.

The total amount of energy in a closed system is always the same. Energy can be changed from one form to another, but all the different forms of energy in a system always add up to the same total amount of energy, no matter how many energy conversions occur.

Law of Universal Gravitation

The law of universal gravitation states that all objects in the universe attract each other by a force called gravity. The size of the force depends on the masses of the objects and the distance between them.

The first part of the law explains why a bowling ball is much harder to lift than a table-tennis ball. Because the bowling ball has a much larger mass than the table-tennis ball, the amount of gravity between the Earth and the bowling ball is greater than the amount of gravity between the Earth and the table-tennis ball.

The second part of the law explains why a satellite can remain in orbit around the Earth. The satellite is carefully placed at a distance great enough to prevent the Earth's gravity from immediately pulling it down but small enough to prevent it from completely escaping the Earth's gravity and wandering off into space.

Newton's Laws of Motion

Newton's first law of motion states that an object at rest remains at rest and an object in motion remains in motion at constant speed and in a straight line unless acted on by an unbalanced force.

The first part of the law explains why a football will remain on a tee until it is kicked off or until a gust of wind blows it off.

The second part of the law explains why a bike's rider will continue moving forward after the bike tire runs into a crack in the sidewalk and the bike comes to an abrupt stop until gravity and the sidewalk stop the rider.

Newton's second law of motion states that the acceleration of an object depends on the mass of the object and the amount of force applied.

The first part of the law explains why the acceleration of a 4 kg bowling ball will be greater than the acceleration of a 6 kg bowling ball if the same force is applied to both.

The second part of the law explains why the acceleration of a bowling ball will be larger if a larger force is applied to it.

The relationship of acceleration (a) to mass (m) and force (F) can be expressed mathematically by the following equation:

$$\text{acceleration} = \frac{\text{force}}{\text{mass}} \quad \text{or} \quad a = \frac{F}{m}$$

This equation is often rearranged to the form:

$$\text{force} = \text{mass} \times \text{acceleration}$$

or

$$F = m \times a$$

Newton's third law of motion states that whenever one object exerts a force on a second object, the second object exerts an equal and opposite force on the first.

This law explains that a runner is able to move forward because of the equal and opposite force the ground exerts on the runner's foot after each step.

Useful Equations

Average speed

$$\text{Average speed} = \frac{\text{total distance}}{\text{total time}}$$

Example: A bicycle messenger traveled a distance of 136 km in 8 hours. What was the messenger's average speed?

$$\frac{136 \text{ km}}{8 \text{ h}} = 17 \text{ km/h}$$

The messenger's average speed was **17 km/h.**

Average acceleration

$$\text{Average acceleration} = \frac{\text{final velocity} - \text{starting velocity}}{\text{time it takes to change velocity}}$$

Example: Calculate the average acceleration of an Olympic 100 m dash sprinter who reaches a velocity of 15 m/s south at the finish line. The race was in a straight line and lasted 10 s.

$$\frac{15 \text{ m/s} - 0 \text{ m/s}}{10 \text{ s}} = 1.5 \text{ m/s/s}$$

The sprinter's average acceleration is **1.5 m/s/s south.**

Net force

Forces in the Same Direction

When forces are in the same direction, add the forces together to determine the net force.

Example: Calculate the net force on a stalled car that is being pushed by two people. One person is pushing with a force of 13 N northwest and the other person is pushing with a force of 8 N in the same direction.

$$13 \text{ N} + 8 \text{ N} = 21 \text{ N}$$

The net force is **21 N northwest.**

Forces in Opposite Directions

When forces are in opposite directions, subtract the smaller force from the larger force to determine the net force.

Net force (cont'd)

Example: Calculate the net force on a rope that is being pulled on each end. One person is pulling on one end of the rope with a force of 12 N south. Another person is pulling on the opposite end of the rope with a force of 7 N north.

$$12 \text{ N} - 7 \text{ N} = 5 \text{ N}$$

The net force is **5 N south.**

Density

$$\text{Density} = \frac{\text{mass}}{\text{volume}}$$

Example: Calculate the density of a sponge with a mass of 10 g and a volume of 40 mL.

$$\frac{10 \text{ g}}{40 \text{ mL}} = 0.25 \text{ g/mL}$$

The density of the sponge is **0.25 g/mL.**

Pressure

Pressure is the force exerted over a given area. The SI unit for pressure is the pascal, which is abbreviated Pa.

$$\text{Pressure} = \frac{\text{force}}{\text{area}}$$

Example: Calculate the pressure of the air in a soccer ball if the air exerts a force of 10 N over an area of 0.5 m^2.

$$\text{Pressure} = \frac{10 \text{ N}}{0.5 \text{ m}^2} = 20 \text{ N/m}^2 = 20 \text{ Pa}$$

The pressure of the air inside of the soccer ball is **20 Pa.**

Concentration

$$\text{Concentration} = \frac{\text{mass of solute}}{\text{volume of solvent}}$$

Example: Calculate the concentration of a solution in which 10 g of sugar is dissolved in 125 mL of water.

$$\frac{10 \text{ g of sugar}}{125 \text{ mL of water}} = 0.08 \text{ g/mL}$$

The concentration of this solution is **0.08 g/mL.**

Properties of Common Minerals

	Mineral	Color	Luster	Streak	Hardness
Silicate Minerals	**Beryl**	deep green, pink, white, bluish green, or light yellow	vitreous	none	7.5–8
	Chlorite	green	vitreous to pearly	pale green	2–2.5
	Garnet	green or red	vitreous	none	6.5–7.5
	Hornblende	dark green, brown, or black	vitreous or silky	none	5–6
	Muscovite	colorless, gray, or brown	vitreous or pearly	white	2–2.5
	Olivine	olive green	vitreous	none	6.5–7
	Orthoclase	colorless, white, pink, or other colors	vitreous to pearly	white or none	6
	Plagioclase	blue gray to white	vitreous	white	6
	Quartz	colorless or white; any color when not pure	vitreous or waxy	white or none	7
Nonsilicate Minerals	**Native Elements**				
	Copper	copper-red	metallic	copper-red	2.5–3
	Diamond	pale yellow or colorless	vitreous	none	10
	Graphite	black to gray	submetallic	black	1–2
	Carbonates				
	Aragonite	colorless, white, or pale yellow	vitreous	white	3.5–4
	Calcite	colorless or white to tan	vitreous	white	3
	Halides				
	Fluorite	light green, yellow, purple, bluish green, or other colors	vitreous	none	4
	Halite	colorless or gray	vitreous	white	2.5–3
	Oxides				
	Hematite	reddish brown to black	metallic to earthy	red to red-brown	5.6–6.5
	Magnetite	iron black	metallic	black	5–6
	Sulfates				
	Anhydrite	colorless, bluish, or violet	vitreous to pearly	white	3–3.5
	Gypsum	white, pink, gray, or colorless	vitreous, pearly, or silky	white	1–2.5
	Sulfides				
	Galena	lead gray	metallic	lead gray to black	2.5
	Pyrite	brassy yellow	metallic	greenish, brownish, or black	6–6.5

Density (g/cm³)	Cleavage, Fracture, Special Properties	Common Uses
2.6–2.8	1 cleavage direction; irregular fracture; some varieties fluoresce in ultraviolet light	gemstones, ore of the metal beryllium
2.6–3.3	1 cleavage direction; irregular fracture	
4.2	no cleavage; conchoidal to splintery fracture	gemstones, abrasives
3.2	2 cleavage directions; hackly to splintery fracture	
2.7–3	1 cleavage direction; irregular fracture	electrical insulation, wallpaper, fireproofing material, lubricant
3.2–3.3	no cleavage; conchoidal fracture	gemstones, casting
2.6	2 cleavage directions; irregular fracture	porcelain
2.6–2.7	2 cleavage directions; irregular fracture	ceramics
2.6	no cleavage; conchoidal fracture	gemstones, concrete, glass, porcelain, sandpaper, lenses
8.9	no cleavage; hackly fracture	wiring, brass, bronze, coins
3.5	4 cleavage directions; irregular to conchoidal fracture	gemstones, drilling
2.3	1 cleavage direction; irregular fracture	pencils, paints, lubricants, batteries
2.95	2 cleavage directions; irregular fracture; reacts with hydrochloric acid	minor source of barium
2.7	3 cleavage directions; irregular fracture; reacts with weak acid, double refraction	cements, soil conditioner, whitewash, construction materials
3.2	4 cleavage directions; irregular fracture; some varieties fluoresce or double refract	hydrochloric acid, steel, glass, fiberglass, pottery, enamel
2.2	3 cleavage directions; splintery to conchoidal fracture; salty taste	tanning hides, fertilizer, salting icy roads, food preservation
5.25	no cleavage; splintery fracture; magnetic when heated	iron ore for steel, gemstones, pigments
5.2	2 cleavage directions; splintery fracture; magnetic	iron ore
2.89–2.98	3 cleavage directions; conchoidal to splintery fracture	soil conditioner, sulfuric acid
2.2–2.4	3 cleavage directions; conchoidal to splintery fracture	plaster of Paris, wallboard, soil conditioner
7.4–7.6	3 cleavage directions; irregular fracture	batteries, paints
5	no cleavage; conchoidal to splintery fracture	dyes, inks, gemstones

Glossary

A

abyssal (uh BIS uhl) **plain** the broad, flat portion of the deep-ocean basin (42)

alluvial (uh LOO vee uhl) **fan** fan-shaped deposits of alluvium that form on dry land (13)

alluvium (uh LOO vee uhm) rock and soil deposited by streams (11)

aquifer (AHK wuh fuhr) a rock layer that stores and allows the flow of ground water (14)

artesian (ahr TEE zhuhn) **spring** a spring that forms where cracks occur naturally in the cap rock and the pressurized water in the aquifer flows through the cracks to the surface (16)

B

benthic environment the ocean floor and all the organisms that live on or in it; also known as the bottom environment (47)

benthos organisms that live on or in the ocean floor (46)

breaker a heightened water wave that begins to tumble downward, or break, upon nearing the shore (80)

C

channel the path a stream follows (7)

cleavage (KLEEV IJ) the tendency of a mineral to break along flat surfaces (127)

condensation the change of state from a gas to a liquid (39)

continental margin the portion of the Earth's surface beneath the ocean that is made of continental crust (42)

continental rise the base of the continental slope (42)

continental shelf the flattest part of the continental margin (42)

continental slope the steepest part of the continental margin (42)

Coriolis (KOHR ee OH lis) **effect** the curving of moving objects from a straight path due to the Earth's rotation (72)

crest the highest point of a wave (78)

D

deep current a streamlike movement of ocean water far below the surface (74)

deep-ocean basin the portion of the Earth's surface beneath the ocean that is made of oceanic crust (42)

delta a fan-shaped deposit of alluvium at the mouth of a stream, where the stream empties into a large body of water (12)

density the amount of matter in a given space; mass per unit volume (125)

deposition the process by which material is dropped or settles (11)

desalination the process of evaporating sea water so that the water and the salt separate (53)

discharge the amount of water a stream or river carries in a given amount of time (7)

divide an area of higher ground that separates drainage basins (6)

drainage basin the land drained by a river system, which includes the main river and all of its tributaries (6)

E

El Niño periodic change in the location of warm and cool surface waters in the Pacific Ocean (77)

erosion the removal and transport of material by wind, water, or ice (4)

evaporation the change of state from a liquid to a vapor (39)

F

flood plain an area along a river formed from sediments deposited by floods (13)

G

gradient a measure of the change in elevation over a certain distance (7)

ground water water that is located within rocks below the Earth's surface (14)

H

hypothesis a possible explanation or answer to a question (112)

I

iceberg a large piece of ice that breaks off an ice shelf and drifts into the ocean (67)

L

load the materials carried in a stream's water (8)

longshore current the movement of water near and parallel to the shoreline (81)

M

mid-ocean ridge a long mountain chain that forms on the ocean floor where tectonic plates pull apart; usually extends along the center of ocean basins (43)

N

neap tides tides with minimum daily tidal range that occur during the first and third quarters of the moon (86)

nekton (NEK TAHN) free-swimming organisms of the ocean (46)

nonpoint-source pollution pollution that comes from many sources and that cannot be traced to specific sites (19, 56)

nonrenewable resource a natural resource that cannot be replaced or that can be replaced only over thousands or millions of years (22, 52)

nonsilicate mineral a mineral that does not contain compounds of silicon and oxygen (126)

O

ocean trench a seemingly bottomless crack in the deep-ocean basin that forms where one oceanic plate is forced underneath a continental plate or another oceanic plate (43)

P

pelagic (pi LAJ ik) **environment** the entire volume of water in the ocean and the marine organisms that live above the ocean floor; also known as the water environment (49)

permeability (PUHR mee uh BIL uh tee) a rock's ability to let water pass through it (14)

petroleum an oily mixture of flammable organic compounds from which liquid fossil fuels and other products are separated; crude oil (52)

plankton microscopic organisms that float at or near the ocean's surface (46)

point-source pollution pollution that comes from one particular source area (19)

porosity (poh RAHS uh tee) the amount of open space between individual rock particles (14)

precipitation solid or liquid water that falls from the air to the Earth (5, 39)

R

recharge zone the ground surface where water enters an aquifer (15)

renewable resource a natural resource that can be used and replaced over a relatively short time (22)

rift valley a valley that forms in a rift zone between diverging tectonic plates (43)

S

salinity a measure of the amount of dissolved salts and other solids in a given amount of liquid (36)

satellite a natural or artificial body that revolves around a planet (45)

scientific method a series of steps that scientists use to answer questions and solve problems (112)

seamount an individual mountain of volcanic material on the abyssal plain (43)

septic tank a large, underground tank that collects and cleans waste water from a household (21)

sewage treatment plant a factory that cleans waste materials out of water that comes from sewers or drains (20)

soil a loose mixture of small mineral fragments and organic material (11)

spring tides tides with maximum daily tidal range that occur during the new and full moons (86)

storm surge a local rise in sea level near the shore that is caused by strong winds from a storm, such as a hurricane (83)

surf the area between the breaker zone and the shore (80)

surface current a streamlike movement of water that occurs at or near the surface of the ocean (71, 73)

swells rolling waves that move in a steady procession across the ocean (81)

T

temperature a measure of how hot (or cold) something is (110)

thermocline a layer of ocean water extending from 300 m below sea level to about 700 m below sea level in which water temperature drops with increased depth faster than it does in other zones of the ocean (37)

tidal bore a body of water that rushes up through a narrow bay, estuary, or river channel during the rise of high tide, causing a very sudden tidal rise (87)

tidal range the difference between levels of ocean water at high tide and low tide (86)

tides daily movements of ocean water that change the level of the ocean's surface (84)

tributary a smaller stream or river that flows into a larger one (6)

trough (trahf) the lowest point of a wave (78)

tsunami a wave that forms when a large volume of ocean water is suddenly moved up or down (82)

U

upwelling a process in which cold, nutrient-rich water from the deep ocean rises to the surface and replaces warm surface water (77)

V

volume the amount of space that something occupies or the amount of space that something contains (121)

W

water cycle the continuous movement of water from water sources into the air, onto land, into and over the ground, and back to the water sources; a cycle that links all of the Earth's solid, liquid, and gaseous water together (4, 39)

watershed the land drained by a river system, which includes the main river and all of its tributaries (6)

water table an underground boundary where the zone of aeration and the zone of saturation meet (14)

wave height the vertical distance between a wave's crest and its trough (78)

wavelength the distance between two adjacent wave crests or wave troughs (78)

wave period the time between the passage of two wave crests (or troughs) at a fixed point (79)

whitecap a white, foaming wave with a very steep crest that breaks in the open ocean before the wave gets close to the shore (81)

Index

A **boldface** number refers to an illustration on that page.

G

H

I

K

L

M

N

O

P

R

S

T

U

V

Index

W

Y

Z

Credits

Abbreviations used: (t) top, (c) center, (b) bottom, (l) left, (r) right, (bkgd) background

ILLUSTRATIONS

All work, unless otherwise noted, contributed by Holt, Rinehart & Winston.

Scope and Sequence: T11, Paul DiMare, T13, Dan Stuckenschneider/Uhl Studios, Inc.

Chapter One: Page 5, Mike Wepplo/Das Group; 6(bc), MapQuest.com; 8, Uhl Studios, Inc.; 11(b), Marty Roper/Planet Rep; 14(cl), Stephen Durke/Washington Artists; 15(c), MapQuest.com; 15(br), Geoff Smith/Scott Hull; 16, Stephen Durke/Washington Artists; 20(b), John Huxtable/Black Creative; 21(cl), John Huxtable/Black Creative; 21(b), Sidney Jablonski; 23(c), MapQuest.com; 26(c), Mike Wepplo/Das Group; 29(cr), Sidney Jablonski.

Chapter Two: Page 35, MapQuest.com; 36(tl), Ross, Culbert and Lavery; 37(tr), MapQuest.com; 37(b), Ross, Culbert and Lavery; 39(b), Mike Wepplo/Das Group; 42–43(b), Uhl Studios, Inc.; 44(b), Uhl Studios, Inc.; 45(br), Ross, Culbert and Lavery; 46, Yuan Lee; 54(tl), Jared Schneidman Design; 57(br), Mark Heine; 63(c), Jared Schneidman Design; 64(c), Bill Mayer; 64(tr), MapQuest.com; 65(tr), Ross, Culbert and Lavery.

Chapter Three: Page 70(tr), Dean Fleming; 71(tr), Stephen Durke/Washington Artists; 71(bl), MapQuest.com; 72, MapQuest.com; 73(c), MapQuest.com; 74, Stephen Durke/Washington Artists; 75, Jared Schneidman Design; 76, MapQuest.com; 78, Jared Schneidman Design; 79, Jared Schneidman Design; 80, Dean Fleming; 82(c), Uhl Studios, Inc.; 83(t), MapQuest.com; 84(c), Marty Roper/Planet Rep; 85, Sidney Jablonski; 86, Sidney Jablonski; 90(br), Dean Fleming; 90(c), Stephen Durke/Washington Artists; 91(c), Marty Roper/Planet Rep; 93(cr), Sidney Jablonski

LabBook: Page 101(tr), Mark Heine; 105(br), Geoff Smith/Scott Hull; 106(b), Mark Heine; 107(b).

Appendix: Page 110(c), Terry Guyer; 114(b), Mark Mille/Sharon Langley; 122, Kristy Sprott; 123, Kristy Sprott.

PHOTOGRAPHY

Cover and Title Page: Yabb Arthus-Bertrand/Corbis

Sam Dudgeon/HRW Photo: p. viii-1, 22(bl), 52, 60(bl), 71, 78, 96(bl), 97(bc), 98(tr, br), 99(tl), 104, 111(br).

Table of Contents: v(tr), Laurence Parent; v(cr), Laurence B. Aiuppy/FPG International; v(b), Victoria Smith/HRW Photo; vi(tl), Rosentiel School of Marine & Atmospheric Sciences, University of Miami; vi(bl), Wayne Lynch/DRK Photo; vii(tr), Mike Bacon/Tom Stack & Associates; vii(b), Norbert Wu.

Scope and Sequence: T8(l), Lee F. Snyder/Photo Researchers, Inc.; T8(r), Stephen Dalton/Photo Researchers, Inc.; T10, E. R. Degginger/Color-Pic, Inc., T12(l), Rob Matheson/The Stock Market

Master Materials List: T25(br), Image ©2001 PhotoDisc

Chapter One: pp. 2-3 Frans Lanting/Minden Pictures; 3 HRW Photo; 4 Tom Bean/DRK Photo; 7(br), Ed Reschke/Peter Arnold, Inc.; 7(bl), Jim Work/Peter Arnold, Inc.; 9(br), Frans Lanting/Minden Pictures; 9(tr), Laurence Parent; 10(tl), The G.R. "Dick" Roberts Photo Library; 10(cl), Galen Rowell/Peter Arnold, Inc.; 12(tl), The Huntington Library/SuperStock; 12(cr), Earth Satellite Corporation/Science Photo Library/Photo Researchers, Inc.; 13(tr), Visuals Unlimited/Martin G. Miller; 13(cr), Earth Satellite Corporation; 17 Rich Reid/Animals Animals/Earth Scenes; 18 Leif Skoogfers/Woodfin Camp & Associates; 19(bl), Laurence B. Aiuppy/FPG International; 19(tr), Wayne Lynch/DRK Photo; 22(tl), Arthus Bertrand/ Explorer/Photo Researchers, Inc.; 24 Victoria Smith/HRW Photo; 27(c), Rich Reid/Animals Animals/Earth Scenes; 28(bl), Donald Nausbaum/Stone; 28(tr), TSA/Tom Stack & Associates; 30 Jeff and Alexa Henry; 31 C.C. Lockwood/DRK Photo.

Chapter Two: pp. 32-33 Stuart Westmorland/Photo Researchers, Inc.; 33 HRW Photo; 34 Tom Van Sant, Geosphere Project/Planetary Visions/Science Photo Library; 38 U.S. Navy; 40(cl), Rosentiel School of Marine and Atmospheric Science, University of Miami; 41(br), James Wilson/Woodfin Camp & Associates; 41(bl), Norbert Wu; 45 NOAA/NSDS; 47(br), Mike Bacon/ Tom Stack & Associates; 47(cr), Jim Zipp/Photo Researchers, Inc.; 48(tl), James B. Wood; 48(cl), Al GIddings/Al Giddings Images; 48(bl), JAMESTEC; 49(tr), E. R. Degginger/Color-Pic, Inc.; 49(cr), Nobert Wu; 50 Paolo Curto/UP/Bruce Coleman, Inc.; 51(br), Bryan and Cherry Alexander Photography; 51(cl), Breg Vaughn/Tom Stack & Associates; 52(cr), TGS-NOPEC Geophysical Company; 53(bl), Institute of Oceanographic Sciences/NERC/Science Photo Library/Photo Researchers, Inc.; 53(bc), Charles D. Winters/Photo Researchers, Inc.; 55(b), E. R. Degginger/Color-Pic, Inc.; 56(cl), Photo Edit; 56(bc), Andy Christiansen/ HRW Photo; 56(br), Ron Chapple/FPG International; 57(tr), Ben Osborne/Stone; 57(bl), Courtesy Mobil; 58 Andy Christiansen/HRW Photo; 59(tl, tr), Courtesy Texas General Land Office Adopt-A-Beach Program; 62 James Wilson/Woodfin Camp & Associates; 66 Gilles Bassignac/Liaison Agency; 67 SuperStock.

Chapter Three: pp. 68-69 O. Brown, R Evans, and M. Carle, University of Miami Rosenstiel School of Marine and Atmospheric Science, Miami, Florida; 69 HRW Photo; 70(cr), Hulton Getty/Liaison Agency; 77 Lacy Atkins/San Francisco Examiner/AP/Wide World Photos; 81(tr), CC Lockwood/Bruce Coleman, Inc.; 81(bl), Darrell Wong/Stone; 81(br), August Upitis/FPG International; 82 Denis J. Sigrist, International Tsunami Center, Honolulu, HI/National Geophysical Data; 87(tl, tr), VOSCAR/The Maine Photographer; 88 Andy Christiansen/HRW Photo; 91(cl), Art Resource, NY; 92(tl), Warren Bolster/ Stone; 92(br), Fred Whitehead/Animals Animals/Earth Scenes; 94(tl), Todd Bigelow/HRW Photo; 94(br), Warren Bolster/Stone; 95(tc), J.A.L. Cooke/Oxford Scientific Films/Animals Animals/Earth Scenes; 95(tr), Visuals Unlimited/ David M. Williams.

LabBook/Appendix: "LabBook Header", "L", Corbis Images; "a", Letraset Phototone; "b", and "B", HRW; "o", and "k", images ©2001 PhotoDisc/HRW; 97(tr), John Langford/HRW Photo; 97(cl), Michelle Bridwell/HRW Photo; 97(br), Image ©2001 PhotoDisc, Inc./HRW; 98(bl), Stephanie Morris/HRW Photo; 98(cl), Victoria Smith/HRW Photo; 99(b) Peter Van Steen; 99(tr), Jana Birchum/HRW Photo; 102 Victoria Smith/HRW Photo; 103 Andy Christiansen/HRW Photo; 107 Victoria Smith/HRW Photo; 111(tr), Peter Van Steen/HRW Photo.

Feature Borders: Unless otherwise noted below, all images copyright ©2001 PhotoDisc/HRW. "Across the Sciences" 66, all images by HRW; "Careers" 94, sand bkgd and Saturn, Corbis Images; DNA, Morgan Cain & Associates; scuba gear, ©1997 Radlund & Associates for Artville; "Eye on the Environment" 31, 67, clouds and sea in bkgd, HRW; bkgd grass, red eyed frog, Corbis Images; hawks, pelican, Animals Animals/Earth Scenes; rat, Visuals Unlimited/John Grelach; endangered flower, Dan Suzio/Photo Researchers, Inc.; "Health Watch" 95, dumbbell, Sam Dudgeon/HRW Photo; aloe vera, EKG, Victoria Smith/HRW Photo; basketball, ©1997 Radlund & Associates for Artville; shoes, bubbles, Greg Geisler; "Weird Science" 30, mite, David Burder/Stone; atom balls, J/B Woolsey Associates; walking stick, turtle, EclectiCollection.

Self-Check

Self-Check Answers

Chapter 1—The Flow of Fresh Water

Page 8: If a river slowed down, the suspended load would be deposited.

Page 12: Answers will vary. A river might slow where there is a bend, where the gradient decreases, or where the river empties into a large body of water.

Page 16: The impermeable rock layer in the aquifer traps the water in the permeable layer below. This creates the pressure needed to form an artesian spring.

Chapter 2—Exploring the Oceans

Page 35: If North America and South America continue to drift westward and Asia continues to drift eastward, the continents will eventually collide on the other side of the Earth.

Page 43: Rift valleys form where tectonic plates pull apart, and ocean trenches form where one oceanic plate is forced underneath a continental plate or another oceanic plate.

Chapter 3—The Movement of Ocean Water

Page 71: Because he was traveling from Peru to a Polynesian island to the west, Heyerdahl would have noticed the wind blowing from the east.